tech talk 2020

A COLLECTION OF THE BEST TECH TALK COLUMNS

TECH TALK 2020

BY

MATHEW DICKERSON

A COLLECTION OF THE
BEST TECH TALK COLUMNS

TECH TALK 2020
A COLLECTION OF THE BEST TECH TALK COLUMNS

ISBN: 9780648782704 (978-0-6487827-0-4)
Library of Congress Control Number: 2020902151
Word Count: 37091
Flesch Reading Ease: 58.3
Flesch-Kincaid Grade Level: 9.2

Publisher: Small Business Ru!es
9 8 7 6 5 4 3 2 1 0

The author can be contacted at ask@techtalk.digital.

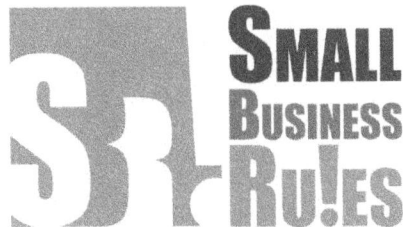

Dedication

In everything I do I have my family at heart. My family members are my inspiration and I value their opinions above all others. It is a dangerous thing to do but with this dedication I want to single out one particular member of my family (at a huge risk of offending the others).

My third book was published in 2009. In my mind I had a plan for more books and the topics and ideas were clearly laid out in front of me. Then a few things got in the way and, before I knew it, ten years had gone by and no further books were produced.

Then my youngest daughter did something special. At the ripe old age of fifteen, she proudly declared that she was going to publish a book of short stories. We all told her that was great and patted her on the back and assumed she would have a different plan next week.

But that didn't happen.

In what seemed like the blink of an eye, my daughter told me she had her book finished and was ready to publish it and wanted to have that completed by her sixteenth birthday. She didn't quite hit her self-imposed deadline but at the mature age of sixteen she published a book of short stories.

That stirred me into action. I told myself that if my youngest daughter could get herself organised enough to publish a book while studying at school and participating in after-school activities and dealing with teenage 'stuff' then surely I could get my act together.

So I did and here I am.

Thank you, Julia, for inspiring me to take action rather than just talk about it. My wife, Katrina, and my three older children, Georgie, Sophia and Andy, have also inspired me as I always want to make them proud of their Dad.

Mathew

CONTENTS

FOREWORD

Tech Talk. That title would be enough for some people to glance at the page and move right along to the next.

Only 'tech heads' or industry people would typically read a column about technical 'stuff', right?

With Mathew's column, you'd be wrong.

As the editor of the Daily Liberal over the period the articles in this book were published, I can tell you they were read in great numbers by our readers.

The numbers from analytics tell us the web versions had good click rates and the correspondence and general comments from the community about something mentioned in one of Mathew's columns showed us his article in the paper was also getting eyes on it.

The reason for this, in my opinion:

Mathew's columns were a bit different from standard tech columns.

They were entertaining *and* informative!

I learnt new things from them, as I am sure other readers did.

Another reason for the success of the column is that Mathew always tackled relevant topics.

Tech Talk covered topics like driverless cars, banks and artificial intelligence with good humour, facts and fun puns, all the while busting a few well-worn tech myths.

I hope you enjoy these columns as much as I did.

Craig Thomson
Editor at Australian Community Media

INTRODUCTION

When you have a passion, any passion, you want everyone else to share in that passion – or at least to vaguely understand that passion. In my early days of writing about IT, stretching way back to 2004, the articles were typically written for IT specific publications and published for an IT audience.

Preaching to the converted would be the expression that comes to mind.

Several years ago in a discussion with Brian O'Flaherty, editor of an Australian Fairfax publication, we talked about the idea of a weekly column that would be of interest to a broader audience, rather than one that was IT focused.

And so *Tech Talk* was born.

After some time honing the style and gauging audience feedback to ensure I was hitting the mark, the decision was made to take the column to a wider audience. After Brian's tragic passing, I worked with Lynn Rayner and Craig Thomson at the local level who then put me in touch with Jen Walker, Group Deputy Editor. Jen had the ability to place the articles in 140 of the Fairfax mastheads across the nation.

After further discussions and additional refinement, *Tech Talk* started to appear in mastheads across Australia and I knew immediately that I was hitting the mark because I started to receive emails from all parts of the country. I created a specific e-mail address for these columns to easily track how the contact was generated and when the questions and comments came in, my first response was always to ask where they had read the article. The answers were often from publications I didn't even know existed! The online versions of these publications also showed good readership statistics to back up my anecdotal information.

Part of Fairfax was sold to Australian Community Media in the middle of 2019 and I now work with Gavin Stone from ACM but the core concept throughout all of these articles is still the same. I want to share my passion for technology with you.

I hope you enjoy reading a year of the progress in technology. And never hesitate to contact me at ask@techtalk.digital.

New Year Predictions

You would think I would be old enough to know better. I have been writing and speaking about technology for over fifteen years and this year is my thirtieth involved in the technology industry yet I still don't seem to have learned from the past.

Every now and then I have a rush of blood after a wild New Year celebration and decide that it is a good idea to make some predictions about what the world of technology will bring this year.

Inevitably when I look back at those predictions twelve months later, I look down at my feet and shuffle them nervously in the hope that everyone was watching the Sydney Test match at the beginning of the year and didn't take too much notice of my predictions.

Despite all of that history, I am going to have another crack this year...Wish me well.

In the world of mobiles, we have witnessed 5G in real-world situations. Across the world, various carriers have started trials or small roll outs of 5G. I have several predictions for 5G for this year. Firstly, we will see 5G available in all major metropolitan areas in Australia but sadly I don't believe we will have 5G in regional inland centres by the end of the year. In terms of devices, I predict that the major manufacturers will have several 5G mobile handsets and mobile broadband devices and notebooks and tablets by the end of the year. Some of these devices will be foldable handsets as foldable phones will make somewhat of a comeback this year – maybe some with foldable screens. While on mobiles, I predict by the end of this year that mobile phone users across the world will be using a combined twenty exabytes of monthly mobile data. An exabyte is a giga gigabyte. A big number! Our current growth rate of mobile data usage globally is growing at over forty per cent. Part of this data trend will be in relation to consuming video content as I predict that officially the Internet will globally surpass TV this year in terms of minutes spent consuming content.

If we look at something else that is mobile – just slightly larger – I predict that we will hit the six million mark for total sales of plug-in electric vehicles by the end of this year. Towards the end of 2018 we hit four million across the world but the pace of adoption is accelerating and I have previously written about the plethora of new electric models that we will see this year. Helping with that prediction is another prediction – Tesla will hit the 500,000 mark for vehicles produced this year. I am not convinced that Tesla will not be bought by someone else by the end of the year but regardless of who owns Tesla, they have to increase their capacity to survive.

While on electric transportation, I predict that major motorcycle manufacturers will introduce mainstream electric models by the end of the year including the most unlikely of them all – Harley Davidson. When you think of a Harley motorcycle you think of throbbing guttural rumbling noise. As the world changes the best companies in the world must also change and I predict that Harley will go along for the ride (I also predict we will see more bad puns in this column this year).

To power these electric cars and motorbikes, we need electricity.

I have two specific predictions for electricity production related to huge projects. I am cheating a little with these two as they are projects that have been announced and approved – but not yet finished.

Firstly, I predict the largest solar park in the world will be built this year in the Sahara Desert. The Benban Solar Park will have the capability of producing approximately 2GW of power. That is larger than the Liddell Power Station in NSW that is being decommissioned in 2022.

I also predict the largest offshore wind farm will be in operation by the end of the year. Hornsea Project One is located one hundred and twenty kilometres off the coast of Yorkshire in the UK. This is a 1.2GW project that will consist of 174 turbines. Hornsea Project Two, Three and Four are also in the planning stages with a total planned capacity of 6GW. The largest wind farm in the world is the Gansu Wind Farm in China with a current capacity of 8GW and a planned total capacity of 20GW.

My final prediction is making predictions in relation to technology is very difficult so take these predictions as a loose estimate and have a wonderful technological year. Journey

THE FUTURE IS HERE NOW

I **did say in my column last week that you would think I would have learned by now that making predictions for the next year in the world of technology is somewhat foolhardy as the technology world changes so quickly.**

I also said that I expect some people to look back at my predictions in a year and have a chuckle about how far wrong some of my predictions were. It didn't take a year. In comments received already, one particular prediction is being ridiculed without even a week passing. The prediction of an electric Harley Davidson motorcycle seemed somewhat unbelievable by some readers.

Time will tell.

While on predictions, the most exciting time of the year is upon us while we speak. The Consumer Electronics Show (CES) has been a forum for manufacturers to demonstrate their latest products for the last fifty years. CES 2019 is in full swing in Las Vegas and, with more than 4,500 exhibiting companies, the show has products that are available now or at some time in the near future. With more than 180,000 attendees at CES, it is paramount that manufacturers bring their latest products to wow the audience.

And latest products they did bring.

I am not convinced we will see some of these products on retail shelves anytime soon but they all exist and were all demonstrated at CES.

The first one that caught my attention was Foldimate. This device folds your clothes perfectly in ten seconds per item. Finally technology that my wife would embrace. Staying in the house, there are lots of ways to add some technology to the lives of your pets. You can have cellular connected collars that let you track and call your dog from your phone, smart doggy doors and facial recognition feeders. One company has released a pet feeder that uses facial recognition to release the correct amount of feed to only the specific pet that requires it. Some may argue that is a waste of technology – but I love the creativity involved even if it does seem like technology for the sake of technology.

Autonomous mowers have been available for some time and, in fact, I have been using one since 2010, but they have improved dramatically. New models are now available that have improved the concept dramatically – and reduced the price.

If you ever wanted headphones that give you electric shock treatment, this is your lucky day! I don't quite understand why but apparently they will give you benefits when learning new skills. I might stick to the old-fashioned methods but maybe they will appeal to some?

If you are a fan of Batman, you will be a fan of accessory belts. Now we have a smart belt that will track your waistline and your activity and take action if you have a fall. No comment on this one!

If you just don't have the time to clean your teeth properly, technology is the answer. A new brush that looks like a mouth guard but it guarantees to give you the best clean your teeth will ever have outside a dental chair – and do it all in ten seconds.

Lastly, what every home needs. A mood lamp that is connected to the virtual assistant, Alexa.

> I have no idea how the mood lamp works or what it does, but connecting it to Alexa must be a good thing?

There were hundreds of other announcements with automated driving and VR entertainment in cars and AR cooking and...the list goes on. Keep an eye out for some of these items and devices at a retail store near you – soonish!

Trust me – I sell technology

Trust in business is an interesting concept. Without thinking too hard about it, we inherently have different levels of trust expectations in various organisations we deal with.

Often it comes down to the importance of what we are purchasing, the price and the length of time we will use it.

When we see a five-dollar ornament at a weekend market, we probably don't think too much about the trust factor. It isn't a lot of money and if the ornament falls apart in a week, the world hasn't come to an end.

There are other organisations where trust plays a much greater role. The medical field springs to mind immediately.

In terms of technology, I have always spoken to my staff in my technology businesses that the level of trust required in the technology world is similar to the trust a business has with their accountant. You pay a sum of money to your accountant that is a reasonable sum of money; the relationship is normally multi-year; you don't want to go to jail but you also don't want to pay more tax than necessary and you want the best advice on how to structure your business.

In the technology world, the sums of money paid to a technology provider are similarly reasonable sums. They may be via a lease in the case of computer hardware for example or an ongoing bill payment with a mobile phone or an outright purchase but whichever way it is purchased, it is a reasonable sum. The relationship is typically multi-year and people want the best advice to deliver a solution and ensure they aren't paying too much.

With all of that in mind, there are several surveys recently released that show the level of trust in a range of technology brands based on the experience that real people have with real products and real services.

I have drawn this information from both locally and overseas.

In one survey of the world's most reputable technology companies, I found some interesting results. At the top was Google followed by Canon; Sony; Microsoft; Nintendo; Intel; Philips; Amazon; Netflix and Samsung. Surprisingly, the highest value company in the world, Apple, was way down the list below other big names like Dell; LG and Cisco. Notice a distinct lack of social media sites in this list – possibly reflecting some of the scandals we saw last year. This type of survey relies on the feel-good nature of the business name and the perception in the wider world when people hear the name mentioned.

The next survey I found related to the culture within an organisation. The top five in this list were Google; Facebook; Microsoft; LinkedIn and Adobe. A common link with Google between culture and reputation. I have always had a belief that if you create a positive culture within an organisation then success is not guaranteed but is much more likely.

If we then start to break down the brands to more specific products, we find more interesting information and possibly even some apparent contradictions. If we look at owner satisfaction with notebooks, Apple is on top of the list followed by Microsoft; Dell; IBM/Lenovo with Toshiba and HP tied. Printer brands show the specialisation in this segment. Fuji-Xerox sits on top of owner satisfaction with Brother up next followed by HP; Canon and Epson. Four of these five brands you would normally associate first and foremost with printing.

For television executives in this country, streaming services are growing at an alarming rate. Users were happiest with Netflix and Apple iTunes Video – keeping in mind that Netflix has eight times the subscriber base compared to Apple – with Google Play; YouTube Premium and Stan following in the satisfaction stakes.

Lastly, I looked at Smart Speakers. Despite the hype associated with Smart

Speakers, their penetration in the marketplace is still only at a low level. It may be a case of the apocryphal story about the two shoe salesmen sent to a developing country. One came home and said, "Nobody wears shoes so there is no market for our product." The other salesman sent a message and said, "Nobody wears shoes so please send some containers of shoes!"

The satisfaction level with Smart Speakers loosely matched the penetration of the brands. Google Home came in first with Sonos One next and the Amazon Echo (and variants) followed by the Apple HomePod.

In technology, there is no doubt trust plays a huge role and if you want to see the brands that will prosper over the next year, the ones mentioned here – with the higher satisfaction and trust levels – will most likely be the brands to prosper.

WRITTEN BY A BOT?

I f you have seen the comedy film, Why Him, you will remember the voice of Kaley Cuoco (famous for her role as Penny in The Big Bang Theory) as Justine, Laird's in-home artificial intelligence.

Justine is incredibly helpful – some might say a little too helpful – and is constantly listening and offering advice as you move throughout Laird's house.

This is a fictional movie so obviously nowhere near the truth.

Hold on for just one second.

Henn na Hotel in Japan employed 243 robots in 2015 to make the hotel the most efficient in the world. That included access to the virtual assistant robot, Churi, in every hotel room. Guests started complaining about the attention though. If you snored while staying in a room you may be woken up in the middle of the night by Churi saying, "Sorry, I couldn't catch that. Could you repeat your request?" Even having a simple conversation in your room with another guest would cause interruptions from Churi.

This is not the hotel to stay in if you are paranoid and think that someone is listening – because they definitely are!

Churi was fired.

Humanoid concierge robots that couldn't answer basic questions were not proving to be very useful.

Fired.

Surely bellhop robots that carried luggage to a set room would be useful? Not when they can only carry luggage on flat surfaces and they could not negotiate their way to every room.

Fired.

In fact, half of the robots were recently fired in sweeping changes throughout the hotel. Luckily there is no robot union otherwise there may have been discussions about retraining and being redeployed elsewhere in the hotel.

Bottom line – they were just not good enough at their jobs.

Is that the end of the robot revolution? Not at all – in fact I see it as just the beginning.

Sure, Henn na Hotel tried an experiment with robots that didn't prove that successful. There always has to be someone who leads the way and breaks new ground so that others may follow. I am sure that right now, new and improved versions of the robots at Henn na Hotel are being tested and modified and refined.

The general public would like to see some robotics. In a recent British survey, 27 per cent of people think a home robot could save them two hours a day. Simple housework – such as cleaning – seems to make sense. I already use a very basic robot at my work and home but it is very much a first revision. Improvements are constantly being made.

In that same survey, 60 per cent of people believe there will be a robot in every home within the next 50 years and only a quarter of them can't imagine a life without a robot as a part of the family.

Hold on.

A part of the family? I will accept – just – a family pet being a part of the family but a robot?

When you also consider that 20 per cent of people say they want a robot to simply keep them company you start to understand the part robots may play in our future lives. It may be scary and it may be fanciful – but more than likely it is going to be reality.

Keep an eye on robot developments. You will see a major advance on the news from time to time but you can be assured that for every one that makes the news with a major breakthrough, there are thousands of other people researching, testing and experimenting to push the boundaries. The next time you visit a motel or speak to someone on the phone, just ask if they are real or a bot. The answer may surprise you.

DRIVERLESS OR DRIVER LESS?

A few years ago I was speaking at a conference near Colombo in Sri Lanka and the conference organiser had booked accommodation for me at the Taj Exotica in Bentota. This was about a thirty-minute drive from the conference centre and each day of the conference the organisers had arranged a car to collect me from my accommodation and drive me to the conference.

The road was supposedly a two-lane road in each direction but the locals had not been informed of this simple fact. Lines on the road meant very little to the drivers there. It was all about the amount of room available on the road.

If a car or bus or bike or Tuk-tuk or donkey could squeeze into a space, it would.

At one stage I counted eight 'vehicles' across the four available lanes. By the last day I asked my driver if I could drive on this road as I thought it was the craziest driving experience I could imagine. By the time we arrived at the conference centre, the assessment given to me by my driver was that I was a 'terrible' driver. He said that I didn't use my horn even once. I needed to use the horn regularly if I wanted to drive on these roads.

With that backdrop, think of just some of the challenges facing car makers as driverless cars are being developed. In the laboratory or on a computer screen the technology might seem to work with no problems at all. Drop that same technology into a real-world environment such as the streets around Colombo and you start to realise just how incredible the human brain is. A human looks at the chaos that is driving and makes a number of different decisions without us even thinking about it. A driverless car has to have all of this information written in lines of code. Given the fact that current technology relies heavily on lines on a road, eight lanes into four is something that a current driverless car would struggle with.

Before we get to that point though, the largest roadblock (sorry – I did promise to reduce those puns this year) is the amount of trust people place in the technology. If you saw a car approaching and you were about to step onto a pedestrian crossing, would you trust that car to stop for you? If a human was behind the wheel, you would typically wait for eye contact to know that the driver was aware of you and then proceed. Where are the eyes of the driverless car? Well, that problem may be about to be somewhat solved.

Land Rover has developed a concept whereby a driverless car will project lights onto the road in front of the car to demonstrate direction and change in speed of the vehicle. A glance at the road in front of a car will show where it is headed and the distance between bars on the road will show if it is about to brake or accelerate. It does sound like a good idea in a pristine environment but I can see a confused number of lights on the road in front of vehicles. At least the manufacturers are thinking about the problem and trying to bring forward solutions.

One BMW board member is not convinced that driverless cars will ever be on the road. The ethical dilemma was cited as the biggest issue that Ian Robertson was concerned about. If a driverless car had to decide between hitting one person or another, how would it decide the course of action?

How does a human decide this same course of action? It would be difficult to think of many scenarios where these decisions are being made and with the inherently safer nature of driverless cars, I would suggest that there will be very few instances where this choice needs to be made.

It may just be a way of BMW trying to turn people away from the technology to cover the fact they are lagging other manufacturers. In relation to whether BMW, or any company, will be able to crack the magic code, I will leave the last words to the world's most famous carmaker, Henry Ford. "Whether you think you can, or you think you can't--you're right."

Salami with that?

Those of you of a similar vintage to myself may remember Richard Pryor acting in Superman III. I remember Richard Pryor more for being a stand-up comedian than an actor in a series of movies starring Christopher Reeve, but I do remember his role in this movie.

Pryor plays the part of Gus Gorman, a computer programmer who embezzled money from his employer using a technique known as salami slicing. The concept is that if you take a very small amount (say, one cent) from a lot of accounts or transactions (say, tens of millions) you can make quite a nice income. There are various incidents in fiction and reality where salami slicing has been the preferred technique of those who like the idea of a get-rich-quick scheme.

You would think that by now, with the sophistication we have in AI and detection techniques, that it would be nigh impossible for someone to get away with a salami slicing scheme.

As much as they will deny it was a deliberate salami slicing technique, Optus has just been fined $10 million over a minor variation of the theme. This technique is called third-party billing.

The idea is that Optus allowed other companies to use the Optus billing system to sell games, ringtones and other digital content and bill the customers directly on their phone bill. Sounds convenient. Unfortunately, some of the third-party companies started billing customers when the customers were not explicitly aware of the fact they were being charged for a service or without realising that clicking on a text may be enough to add a small component to their bill.

The amounts being charged were typically small. Less than $3 for most of the services. Some of the billing providers hoped that, in the whole scheme of things, customers would not notice an extra $3 on their phone bill that may have totalled several hundred dollars. As with the salami slicing technique, it soon adds up. Optus customers were charged $195 million from 2012 to 2017. Some customers may have been aware they were being charged but the ACCC claimed Optus did not properly inform customers about a variety of aspects of the charges.

Why would Optus allow this to happen to their loyal customers? A small matter of $66 million in commissions earned through the process probably helped Optus glance the other way. So far Optus has refunded 240,000 customers a total of $8 million but that may just be the beginning. Optus received a total of 600,000 enquiries from customers over the five-year period.

The lesson here for the consumer? Check your bills. Not just your telecommunications bill but all of your bills. Your electricity, your gas, your rates. It may be a simple mistake or a disguised salami slicing scheme but it is your money. Find the time and have a look. You might save enough to buy more salami!

ELECTRIC MAIL

In my youth growing up in Dubbo, one of my favourite pursuits was motocross racing at Morris Park – and other venues around the State. I remember one particular rider who was an exceptional racer. The story around the track was that he honed his skills during the week doing his day job. He was a postie.

He had his faithful Honda CT90 that he would ride up gutters and weave in and out of trees and the occasional traffic. At the time I thought it sounded like a great job. Ride a motorbike all week and then race on weekends!

Well for a modern postie, that world is about to change.

With electric cars slowly making inroads (tick for weekly pun) it is inevitable that electric motorbikes will also start to charge ahead (too many puns?) For the modern postie, the change is slightly different. An electric motor will replace the internal combustion engine but the transport method will also grow by one wheel. The bike will turn into a trike.

At first, I thought this may have been related to the actual size of the battery required. But no. Back in the seventies when the CT90 was the main delivery vehicle, letters were being delivered. This is in the days before e-mail when people put pen to paper. But now Australia Post is increasingly being used to deliver parcels. The volume of parcels through Australia Post has grown ten per cent per year over the last three years and the expectation is that by 2020 one in every ten retail items purchased will be bought online. Tough time to be a retailer!

In addition to the e-trikes, Australia Post has ordered a new fleet of 4,000 electric pushbikes. I first rode an electric pushbike at a conference in Tasmania four years ago and they are an impressive piece of engineering. They are designed to add to your pedal power rather than take over from it – it is not designed to be a motorbike.

The added bonus for the e-trike delivery drivers (not sure if they are drivers or riders) is that it keeps them safer. Not necessarily from falling from their bike – but from the sun and from swooping magpies as the trikes have a small canopy. A trial has been running in Tasmania since 2017 but expect to see these e-trikes on our roads in the area soon.

The only down side I can see with the change in machinery is that any budding motocross riders may no longer be able to hone those skills for the track – but who knows. How long before we will start to see e-trike racing? We could even attend without needing to pack our earplugs.

CLEANED OUT

Back in my days of running my Managed Service IT business where we managed hundreds of computer networks for clients across the nation, I used to tell a story of the importance of holistic security.

We would setup servers and routers behind firewalls and keep the systems updated and patched and as secure as possible within the budget of the client. All of this was to protect client networks from attacks via the Internet or internally or with spoofed Wi-Fi.

But then I read about an incident in the US where a small group of men posed as a cleaning crew and entered an office during the late afternoon. They went about their cleaning activities raising no suspicions.

When all the employees went home, the 'cleaners' entered the server room and physically removed all the servers from the server room and left the building with them!

The hacking attempts on Australia's major political parties is one that should send shivers down the spines of all Australians as a potential attack on our democracy.

It is generally accepted that the US has Trump instead of Hillary as President due to hacking of Democratic Servers and 'spear fishing' e-mails. This then allowed a systematic release of false information combined with confidential documents into the American information ecosystem in the lead up to the election.

There are a number of methods that hackers use to interrupt normal operations for a business or political party and many are much easier than gaining access to the actual servers.

Denial of Service (DoS) attacks can flood a server with so much information that it renders it all but useless. Staff could leave a notebook in a café while they pay a bill and cookies (the electronic kind) can be stolen before the staffer returns – which would effectively hand over a number of passwords and usernames. E-mails can be sent with 'urgent' information that needs to be opened – which installs keylogger software which sends a hacker every keystroke typed in to a PC. During the course of just a few days, that would effectively include most passwords used to gain access to resources on a server.

In the movie about his hacking activities, Edward Snowden would always cover the camera and microphone on his PC to physically prevent eavesdropping. As much as we think of hackers running brute-force algorithms and lines of text whizzing down a computer screen, it is sometimes the simple items that users forget. How many homes have smart TVs and smart home devices such as an Amazon Echo or a Google Home Mini? I am sure hackers would find it easier to eavesdrop via some of these devices to a home of a government employee than target a server of the Australian Government.

There are approximately 14,485 nuclear warheads in the world but the next world war may not be fought on battlegrounds by troops with guns.

Governments are employing skilled computer technicians to attack the networks that make societies function. Keep that in mind when you next type in a password of 12345!

THE KING IS DEAD. LONG LIVE THE KING!

When I first started selling mobiles phones in 1990, there was no doubt about the number one brand. Motorola. If you wanted to sell mobiles but didn't stock Motorola, then you weren't serious. Motorola had a proud heritage in mobiles with the first ever mobile call made from a Motorola phone back in 1973 and the terms 'brick phone' and 'bag phone' referred specifically to two Motorola models.

But the problem with being the incumbent leader is that market leadership can turn on a dime at the smallest hint of complacency. Motorola was a proud American company but a company in the relatively tiny country of Finland, a country that is one sixtieth the size of the US, came along and stole the market away from Motorola. By the late 1990s and early 2000s, Nokia owned the worldwide mobile market. To this day, Nokia has seven of the top ten best selling phone models ever.

Then Apple changed the face of mobiles. The iPhone launch, on 29 June 2007, forever changed what consumers expected from their mobiles.

Nokia was left behind and we had a new king. If history has taught us anything, it is that a king is not forever.

A combination of innovation and complacency drive the mobile phone market. This week we saw an unprecedented event. Samsung held a regional sales training and launch event for their newest phones on the market. The Samsung Galaxy

S10; S10+ and S10e were all on show in Dubbo this week along with a discussion on the next greatest innovation we will see in the mobile market, the Galaxy Fold.

Samsung are after the king's crown – and to do that they know they need to out-innovate and out-market the current king. Not easy when the current king is known to be a leader in both areas. Samsung have managed to crack a major innovation in their latest S10. Fingerprint reading through the face of the screen. In the past, phone manufacturers have placed a dedicated fingerprint reader at the base, side or back of a phone. With the new ultrasonic fingerprint reader, Samsung users have the ability to place their finger on the face of the phone to unlock the device. No dedicated reader. This is a major technological advancement.

Of more importance though is the Galaxy Fold. In folded mode, it shows a still reasonable 4.6-inch display. But unfold your phone like a book and you are presented with a 7.3-inch display – similar to a small tablet.

Despite assurances from Samsung that they have tested the hinge over 100,000 folds, consumers will still want to see how it performs in the real world.

For the marketing value though, these two innovations alone put Samsung in a new light. Are we seeing the changing of the guard? Is the king being dethroned as we speak? Only time will tell but one thing is certain. Incumbency and complacency deliver poorer results for consumers than competition and innovation.

FASTER THAN A KIWI

The most exciting technological aspect of the 2007 federal election was the promise of a 'super fast' national broadband network (NBN). The initial proposal was to see 98 per cent of Australian households connected to the new technology. Over the last twelve years a lot has happened. One significant change in government saw a disappointing change in direction for the NBN with a hybrid model that was designed to save money.

While we can argue the merits of this change, the latest report from the global rankings company, Ookla, shows Australia in a disappointing light.

Australia has fallen to 60th place in download speeds.

That is behind a raft of developing nations but most disappointing is that we are now behind New Zealand. We can't beat them at Rugby but at least we used to have better Internet speeds... until now.

Australia's average fixed line speed was just over 33Mbps and the average upload was 13Mbps. To put these numbers in perspective, Singapore topped the list with download speeds of 197Mbps. We are now behind nations like Malta; Andorra; Qatar and Trinidad and Tobago. We have slipped five places from last year when we were in 55th place and a separate study in 2017 had us at 50th place.

If you contrast this with our mobile speeds, we are in 6th position with average download speeds of 56Mbps and upload speeds of 13Mbps. With our mobile providers just starting to roll out 5G we can expect to see these speeds increase even further and push Australia even higher.

What is hard to fathom for many people (including me) is that the mobile speed is 70 per cent faster than the fixed line speeds. There has already been significant discussion in the industry around how much the latest mobile technologies will start to erode the penetration of the NBN. With many users who only require lower amounts of data already using mobile data in preference to fixed-line connections, the advent of 5G and better mobile coverage will see even more users turn off the NBN altogether.

The NBN does have two tricks up its sleeve though. Firstly, anyone that has high data usage requirements (translate: kids) will still want to use NBN with its cheaper pricing for large data amounts. Secondly, in particular with the premises connected with fibre, it is relatively easy for the NBN to turn up the speed. The technology is already there to deliver 1000Mbps for example, so surely we will start to see more options along these lines to keep the NBN competitive against the 5G threat.

Whichever way we look at it, we need to increase the speed and reliability of the current model NBN to stay competitive in the world – and at least beat our Kiwi counterparts at something.

CUTTING YOUR TIES

How are those Russian election conspiracy theories going? First, in the middle of a NSW State election, the Electoral Commission computers go down which almost brought pre-polling to a shuddering halt.

This computer network has some additional complexity due to the nature of building a huge network across the State for a short period once every couple of years but they know about the next election years in advance so there is really little excuse for not having it working and tested and ready to operate. That seems like a little one-off inconvenience.

Then, to add some fuel to the fire, Facebook goes down. Facebook does what? Facebook is admittedly a huge connected network with 2.32 billion active users across the world but it seems inconceivable that their network would go down.

OK, so I am not thinking that the Russians or Chinese are hacking the Electoral Commission computer network AND Facebook to impact the NSW election but it does highlight how much we rely on so many things that are completely outside of our control.

Think of everyday items we come to use and rely on. Electricity; telephones; gas; water; petrol; food. So many items have such a complex supply chain that an interruption in one part of the world can have far-reaching effects.

I think we need a modern equivalent of the famous quote from sixty years ago by Edward Lorenz, "Does the flap of a butterfly's wings in Brazil set off a tornado in Texas?"

Perhaps a modern equivalent: "Does a cable cut in a data centre in Reno cause a café to go broke in Sydney?"

Just as with Lorenz's butterfly effect, the consequences seem to far outweigh the initial action but we do live in a highly connected world. When that café tries to charge someone for their coffee, does the owner of that café realise where the data for their Point of Sale (POS) system is stored? What connections are in place between the café and their data. They are relying on power supplied by a variety of organisations to complete the link.

A number of Internet Service Providers (ISPs) will provide data linkages from the café to their actual POS data stored somewhere in the cloud. That is just for the transaction. Then their EFTPOS machine will require an entirely different set of connections to be able to process the monetary transaction. It may be a little far-fetched – as is the original butterfly notion – but most organisations across the world right now are relying on linkages, companies and connections that they have never heard of and are completely unaware of.

What can we do about it? Go off the grid and revert to paper and pencil? For a technology column, I am unlikely to suggest that. What I do suggest is to be vaguely aware of what connections we have and, if given a choice, choose providers a little closer to home to reduce the number of steps and potential disconnection. In the meantime, embrace some time without Facebook.

TOO MANY PIXELS?

Have you ever had that feeling of extreme excitement when you have some wonderful news to tell everyone – but it just doesn't seem to register above zero on the excitement scale for anyone else?

I suspect that Samsung is feeling that way right now.

In my Tech Column 124 I wrote in detail about the limits of the human eye related to resolutions of TV screens. The summarised conclusion was that a 4K TV at 75" in size would require a human at closer than two metres to spot a pixel. Jump to an 88" TV at 8K and you would need to be viewing it from just over a metre to spot a pixel. I'm not sure about you but I tend to sit more than a metre from my TV.

So you can feel for Samsung. Next week they launch a range of 8K televisions in Australia. They will have a 65" model for $9,999; a 75" for $12,999K and an 85" for $17,499. A 98" will launch later this year and I won't tell you the price – suffice to say the price tag will be designed for people who don't look at price tags!

Based on my previous column in relation to the limits of the human eye, I am not sure it is even worth considering anything below the 85". Then comes the second part of the problem for Samsung (and Sony and LG who are also launching 8K TVs).

Content.

Most current TV viewing is in High Definition (HD) with some providers boasting about their wonderful new 4K content. Foxtel has started limited sports broadcasts in 4K and some streaming services have limited content in 4K. Put 4K content onto an 8K TV and it will probably look pretty much like it does on a 4K

TV. None of this logic means the manufacturers won't continue though.

We are talking about technology – where too much of anything is never enough.

To be fair to the TV manufacturers, providers are looking at 8K content. Japanese broadcaster, NHK, started broadcasting limited content in 8K at the end of last year and streaming providers are looking at how they may broadcast in 8K in the future.

Which then brings us to the next problem.

Download speeds. Streaming 8K content would require a bare minimum of a connection capable of consistent 50Mbps download speeds. As the various versions of the NBN roll out across the nation, not all of the NBN options are capable of this.

So, apart from the limits of the human eye, the price tag, the lack of content and poor Internet connectivity, I can see 8K TVs being a big hit!

Have a look at them at your favourite electronic goods retailer from next week.

LOOK OVER MY SHOULDER

O ne of my greatest disappointments in society is a syndrome I call LOMS. Look Over My Shoulder. Some people struggle with someone else, who they see when they look 'over their shoulder', gaining some perceived advantage or accessing something they want.

It is an idea not limited to individuals. Businesses do it as well. Even large businesses. Very large businesses. Take Google (or its now parent company Alphabet Inc.). Current market capitalisation is $818 billion. For the last five years, it has sat comfortably in the top four stocks on the US Stock Exchange for total market capitalisation. Its business model is stable and it has 89.95 per cent market share. Revenues are consistent and continue to develop.

But Google did a LOMS and saw Facebook and Twitter and wanted to own that space as well. They tried several times. Have you heard of Google Orkut? Google Friend Connect? Google Buzz? Most people haven't but, after three failed attempts, Google+ was going to be the social network to end all social networks. It was going to be bigger than Facebook and Twitter. Google+ introduced 'Hangouts' which was a way to video chat with multiple people. It tried features that were designed to compete with Facebook but do it better. When it appeared that things weren't going so well, Google decided to force Google+ onto you. If you wanted Gmail or a YouTube account or just about any Google service, you needed to be a part of the Google+ ecosystem.

Great for user numbers and stats on reports but not great for consumers who want ease of use.

Google also made you use a real name. Wow! There goes half the Facebook audience out the door!

So audience numbers were not great. Engagement was low. The average user session was less than five seconds. And then a security breach.

Google noted that it discovered and fixed the bug in March last year but are just informing users about it now. The bug gave apps the ability to access data that users marked as private. Information such as names; e-mail addresses; gender; age etc. Google said that profiles of up to 500,000 Google+ accounts were potentially impacted but the number of people actually impacted is likely minimal. Given the 300 million users on Google+, the 0.17 per cent of users potentially impacted seems small but I suspect it was the straw that broke the camel's back. Google+, very simply, was just not going to take over from Facebook, so the decision was made to shut it down.

Shutting down a social network can be almost as complicated as starting a social network. All those photos that people have uploaded? How do they access those? Comments on YouTube videos? It is probably about now that some Google execs are questioning the logic of starting a Facebook competitor in the first place. And there is probably a lesson in there for all of us. Do what we do well – and allow others to do the same!

Busting electric myths

Time for me to do my best impression of Adam Savage. I want to bust a few myths today. There seems to be so much discussion around the technology associated with electric cars at the moment – and I struggle to find data and facts in amongst the spin.

I am not formally qualified to add research data to this debate but I feel I have enough experience with ground truths to add some common-sense and critical thinking. From an early age the concept was always instilled in me to use utilise research and data rather than fanciful opinion when it comes to presenting a point of view but I also have a long history with electric vehicles.

I bought my first hybrid fifteen years ago and I have used a total of seven fully electric or electric hybrid vehicles as my main vehicle in the ensuing period. On 18 August 2005, a former Fairfax editor, Linton Besser, wrote a story on my push to have Dubbo City Council use hybrid vehicles and I was the first Mayor in the nation to use an electric vehicle as the main Mayoral vehicle. I currently drive a Tesla which I purchased in March last year and it now has over 40,000km on the odometer.

So with that background, we should explore some of the wild statements being made by people either displaying ignorance or with vested interests.

Myth 1: Charge time.

I have heard anywhere from 5 minutes to several days quoted. There are a variety of types of chargers and companies providing charging infrastructure. Tesla opened a Supercharger station this week in Dubbo and they are the best known with 12,888 Superchargers across the world. Tesla chargers will only charge Tesla cars but there are a variety of universal charger networks that will charge all electric vehicles (including Teslas) with appropriate adapters.

You have different connectors – such as CHAdeMO and J1772 and Type 2 and Tesla – and different suppliers of the infrastructure – such as Tesla and NRMA (one to be opened in Dubbo shortly) and Chargefox and more. Some are free to charge and some charge a rate that is comparable to your home electricity costs.

The all-important question though – how fast?

Just like your mobile phone, the charging rate slows as it nears full capacity but my car will charge at a Supercharger at 600km/hour. I pay nothing to charge at a Supercharger. At home I charge at 80km/hour and I use solar panels to charge BUT if I was paying for electricity, it would cost about $15 to 'fill up.' New V3 Superchargers are being rolled out by Tesla which will charge at 1,500km/hour.

Myth 2: Range.

One of the issues with some of the early electric vehicles was range. An earlier electric car I drove had a range of 170km. Perfect for my job in attending to events in Dubbo but not great if I needed to drive to Sydney daily.

A reduction in battery price and increased efficiency means that the range of electric vehicles is constantly improving.

My current car has a range of 632km on a single charge – even further if I drive slowly. The average Australian drives 15,530km per year or just under 300km per week. Many reasonable priced new electric cars are coming out with a range of over 270km per charge. For most people they can then charge them once or twice a week. There are also many examples of people that have driven around Australia in electric vehicles. Charging stations are typically 200km apart.

On a long trip you can drive for 2 hours and have a 20-minute break while topping up. There are currently about 6,400 petrol stations across this nation but only about 1,000 electric charging stations. It is early days and that ratio will change. I think you will start to see some of the petrol stations adding electric charging stations as it is a logical step for them. This lack of charging infrastructure is a current weakness but one that will be addressed over time.

Myth 3: Cost.

Many people point to a Tesla Model S – with a starting price of $120K – as the prohibitive factor in purchasing an electric vehicle. There will be at least eight all electric cars available in Australia by the end of the year and some will start below $50,000.

That may still seem expensive, but think of the total cost of ownership.

Over a five-year period, the cost of owning a vehicle is the purchase price minus the resale price plus the ongoing fuel and maintenance and insurance/registration costs. At the average distance travelled by a vehicle with average fuel economy and average petrol prices with CPI built in and adding in regular maintenance, an internal combustion engine (ICE) car would cost almost $16,000 in ongoing costs excluding insurance and registration.

There is more to wear out on an ICE car as well so the resale value of a car with almost 80,000km on the odometer would be less. An electric car, assuming similar insurance and registration costs, would reduce the $16,000 running costs dramatically and would lose fewer dollars in resale value. Suddenly the $50,000 electric vehicle stacks up quite well against a $30,000 ICE vehicle.

Myth 4: Utes and performance.

I was once a hot-headed young twenty-something with my V8 253 Holden HZ Ute with a modified exhaust that probably damaged my hearing when I put my foot down. The Toyota Hilux ute has been quoted as the most popular vehicle in Australia and the accusation is that if we go to electric vehicles it will destroy modes of transport for tradies and for people who want to have weekend fun!

How ridiculous.

My old ute could do 0-100 in 9.1 seconds. The Toyota Hilux struggles to get below 10 seconds. One of my previous electric cars was a Nissan Leaf which was purchased for under $30,000 and it had 0-100 time of less than 8 seconds. The

Tesla Model S can do 0-100 in 2.6 seconds.

There are people in our nation right now who have done manual conversions of their Toyota Hilux from ICE to electric. The performance is fantastic and the range is only limited by the amount of battery power provided. Electric cars can have towbars and can be just as much fun as an ICE.

This is probably one of the silliest arguments put forward so far.

Myth 5: Percentage of sales.

I do some talks for different groups wearing the hat of a futurist. In those talks I predict that, regardless of government intervention, 20 per cent of new car sales across the world will be electric cars by 2025. In the current debate, there has been a target of 50 per cent of all new car sales to be electric by 2030. Both figures are eminently achievable. Across the world, new car sales for 2018 show fascinating results.

Norway already sits at 49.1 percent. Iceland is at 19 per cent. Sweden and the Netherlands are at 8.2 and 6.5 per cent respectively.

While the US is only at 2.1 per cent, California has specific incentives to reduce air pollution and, with a population of 40 million, sits at 7.8 per cent of new car sales. Australia, through a distinct lack of leadership, sits at 0.2 per cent. This is one of the lowest in the world.

That will soon start to change.

Myth 6: The electrical network.

The argument is that if everyone changed from ICE to electric cars, the network couldn't handle the load.

We need to look at the numbers.

There are 1.1 million new cars sold in Australia each year. If 20 percent were electric, that would be 220,000 in one year. Electric cars use about 16kWh per 100km. At the average distance driven, 220,000 cars would add up to 3.4 billion kilometres driven. That is a total of 512GWh of electricity.

In comparison to Australia's total electricity consumption, that is in the vicinity of 0.22 per cent. Put another way, the 33 wind turbines at Bodangora produce 0.19 per cent of the electricity used in Australia.

We have six years before 2025 when we will hit 20 per cent of new car sales and we will only need to add 0.22 per cent to total production of electricity of each year we hit our 20 per cent.

It will be some time before all 19.2 million cars on our roads are electric but even if every single car magically transformed to electric tomorrow, total electricity consumption would add less than 20 per cent to our current usage.

Don't forget that oil refineries use power as well so if all oil refineries stopped producing oil tomorrow, there would be more power available, which leads to my next myth.

Myth 7: Power consumption of refineries.

It takes 7kWh to refine 5 litres of petrol for an ICE to travel 47km. An electric car can travel 44km on 7kWh. Stop refining petrol and the grid will have all the power you need. I have researched the data on this from many angles and, although the 7kWh for the refining of 5 litres of petrol is about right, the 7kWh does not all come from the grid. There is power used by a refinery that comes from the crude itself and transportation that is not from the grid and losses in energy efficiency so a more accurate figure might be 1kWh of actual electricity provided by the grid for 5 litres. A more accurate comparison might be the energy saved by not refining oil might translate to 6km of travel not 44km of travel. A saving nonetheless but not a complete replacement.

These are the most common myths that have been thrown around. Look on some EV forums or talk to some EV owners and find out the facts to satisfy yourself that an EV world is the world we will end up with – not necessarily through government intervention but by simple market demand.

WHEN IS A WATCH TOO SMART?

One of the greatest challenges for lawmakers across the world is keeping up with changes in technology. Nowhere in the job description of a tech innovator does it mention that they should hold back on new ideas until legislation can be put in place.

In the highly competitive tech world, new features and devices are brought to market as quickly as possible. Then it is just a matter of waiting for the law to catch up.

This isn't a new scenario.

When Karl Benz received his motorcar patent in 1886, the list of road rules would have been fairly limited.

Once cars started appearing beside horses and bicycles, legislation had to quickly catch up.

Even think back to 1983 when shearers wanted to go past a 2.5-inch comb on their shearing handpiece. This new technology needed legislators to step in to legalise the devices.

The latest challenge for lawmakers relates to smartwatches. A woman in Australia has been charged with murder substantially based on the data on her mother-in-law's smartwatch. The watch showed an activity timeline in contradiction with the daughter-in-law's detailed evidence. Solicitors for the daughter-in-law are arguing that the smartwatch data cannot be relied upon. In a previous case, a Canadian law firm used data from a fitness tracker to show that a person involved in a personal injury lawsuit had significantly reduced their activity after an accident. This objective measure of movement is seen as more reliable than "I don't feel like walking out the front door to collect my newspaper." In another case, a woman in the US was charged with filing a false report when she reported she was asleep in her bed when awoken and sexually assaulted. Her smartwatch data contradicted her stated evidence.

These may be extreme examples and I am sure the vast majority of readers are not about to change murder plans to account for someone wearing a smartwatch but many of us use our smartwatches in everyday life.

The first car I had access to drive as a teenager was my brother's Valiant and the only music in that was the whistle of air when I wound down the window for my crude form of air conditioning. When my brother went all out and installed a radio, it was a pretty big thrill to wind the dial between the two radio stations available. I am sure lining up the little needle on the dial reduced my driving capabilities but there was no law that said I couldn't change radio stations. Across the world, different laws apply to smartwatches and driving. Some jurisdictions define a smartwatch as a Visual Display Unit (VDU) and it is illegal to look at it. So looking at the time could land you a fine. Others define the legality based on whether your smartwatch is actually a standalone phone and still others cover a smartwatch in the general category of 'careless driving' which leaves it as a subjective decision in relation to legality.

Broader questions about data and storage also come into play. When you fill in your life insurance form and tick the box to say 'moderate' physical activity, how long will it be before insurance companies will demand activity logs from the last three months from your smartwatch? Will smartwatch companies start to sell data so that those with reduced physical activity start receiving ads for gym membership! Where will it all end?

Have a great Easter break and feel comfortable in the fact that, for the moment at least, your smartwatch is not tracking how many Easter Eggs you consume over the weekend!

TAKING A BITE OUT OF THE BANKS

Now we tend to think of our big four banks as, well, big. CBA; Westpac; ANZ and the NAB sit at numbers one; three; five and six respectively on our ASX200 by market capitalisation.

And then there is Apple.

To give you some idea of size, their market capitalisation is currently 3.4 times the combined market cap of our big four banks combined. Or to put it another way, Apple alone is the equivalent size of the combined largest 56 companies on the Australian ASX200.

OK – we have established the size of Apple. But with that size comes power.

Way back in 2016, I wrote a column that discussed the ongoing technology battle in Australia between Apple and a group of our big banks. The battle all relates to Apple Pay. On 9 September 2014, Apple announced that they were launching Apple Pay. It gave users of certain models of Apple phones and watches the ability to use their Apple device to make a payment instead of using a contactless card. Contactless cards were relatively widespread by 2014 as they had started being used in 2008.

That all sounded wonderful but, of course, Apple wanted a little slice of the interchange fees (or merchant fees). In Australia, they add up to almost $3 billion annually.

The big four banks in Australia are accustomed to being able to flex their individual or collective muscles and get what they want. We have already established the relative size of Apple so when Apple knocked on Australia's door in 2016, Apple dictated the terms. The banks didn't like that so CBA; Westpac; NAB and Bendigo and Adelaide Bank put a joint submission to the Australian Competition and Consumer Commission (ACCC) to allow them to take on Apple as a group.

ANZ didn't join the joint action. On 28 April 2016, they announced that ANZ users could start to use Apple Pay.

Almost immediately, online credit card applications for ANZ increased by twenty per cent and traffic to their Web site increased by six per cent. Within four months, over a quarter of a million ANZ customers were using Apple Pay.

It must have been killing the other big banks seeing them bleed customers to ANZ while they waited for a determination by the ACCC.

On 31 March 2017 the ACCC announced that they were denying the application by the four banks to collectively boycott and bargain with Apple Pay. "Damn!" said the banks. "Yay!" said Apple. "Who cares?" said consumers." If we want to use Apple Pay, we will just apply for an ANZ card."

As much as the other banks tried to hold out on Apple, surveys and feedback from customers kept mounting the pressure on them. The CBA was the first to give in. On 23 January this year, CBA customers could finally start using Apple Pay. The details are confidential, of course, but I can guarantee that the agreement would not have been as favourable as the ANZ deal. Westpac and the NAB continue to hold out on Apple but mounting pressure from customers and shareholders will see them finally come to the party.

Australians love to adopt new technology. Australian bank data shows that 74 per cent of all MasterCard in-store transactions are now contactless and that per capita, contactless payments in Australia are amongst the highest in the world. The evidence from ANZ shows that once the technology is available, people will adopt it. It won't be a matter of if but when the rest of the banks join ANZ and CBA in offering Apple Pay to their customers.

DENSE COMMUNICATION SPEEDS

Well I have good news and bad news. In the lead-up to the 2007 Federal Election, we heard the vision to commence construction of a super-fast National Broadband Network (NBN) as Australia was lagging behind the rest of the world in Internet connectivity. The original plan was to have fibre to the premises (FTTP) technology for 98 per cent of Australian households.

Finally some recognition that we need alternatives in this nation when we run out of metals and minerals to dig up.

The NBN has indeed been delivering increased speeds to residents across the nation...but...it is not progressing as quickly as other nations.

> The watering down of the technology solution by subsequent Governments meant that we haven't received the full delivery of what was initially promised.

In the latest global rankings report by Ookla, Australia has dropped to number 62 in the ranking of household speeds. On the plus side, over the last year we have improved our average download speed by 28 per cent but we have dropped from number 55 in the world a further 7 places. That puts us behind other global Internet heavyweights (if I had one, I would insert a cute sarcasm emoji here!) such as Kazakhstan and Trinidad and Tobago. More importantly, it puts us way behind New Zealand who sit at number 23. Constantly beating us at Rugby is one thing but having average download speeds that are two and a half times faster than us is just one step too far. To put it further in perspective, our average speed of 35.11 Mbps is bettered by Singapore, who top the latest list, by a multiple of 5.7 times. Now I can understand that Singapore is a more densely populated area than Australia as they sit at number two in the world in population density, so there may be a reasonable argument for that differential – until I discuss the next list from Ookla.

Despite this disappointing and bad news (I know the bit about our speed improving seems like good news but it is just not enough) there is some good news.

When you look at the list of download speeds for Mobile Internet rather than Fixed Internet, we fare dramatically better. Top fifty I hear you ask? Could we dare to dream of top twenty or maybe, just maybe, beat New Zealand?

In Mobile Internet download speeds, we sit comfortably at number five – with an average speed of 58.87Mbps. The top performer in this category is Norway at 67.54Mbps so we are only 12.8 per cent behind the best in the world and while New Zealand sits in the top twenty, they are behind us!

Suddenly the argument about a sparse population doesn't make sense. We are a sparsely populated nation – sitting at number 192 in the world – but our Mobile Internet speed seems at odds with our Fixed Internet speeds. Even more amazing is the fact that our average Mobile download speed is 68 per cent faster than the Fixed speed.

You can understand why NBN is worried about Mobile only households.

Currently, 15 per cent of users across the nation that have access to the NBN choose to go Mobile instead but with the advent of 5G, the estimations are that this figure will soon move to 30 per cent. Every new Mobile connection makes the business model of the NBN that little bit tougher.

NBN will be favoured by users who have high data consumption but the challenge for the NBN is to deliver Fixed speeds that make the NBN compelling to connect to.

Tell me what you think of your Internet speed at ask@techtalk.digital

SOMEBODY'S WATCHING YOU

In my formative years, I learned that a true test of character is how someone behaves when no one is watching them. It would appear that lesson was not widely distributed because the latest data on dashcams is quite interesting.

More on that in minute.

As so often happens with technology, there was an appetite for dashcams well before the technology could support it. We first started seeing the concept of a dashcam during car races. The world famous Hardie-Ferodo 1000 at Mount Panorama in 1979 was the first time Australian audiences witnessed footage from within a car. With a huge studio camera adapted for a car and a three-point mounting bracket from the roll cage, it wasn't something that immediately seemed obvious would one day translate to the modern dashcam.

Texan Police were the first to latch onto the idea of using a camera to capture evidential information. They took a studio camera and connected it to a VHS recorder in the car and mounted it on the dashboard. Fairly crude but somewhat effective.

As technology progressed, three significant changes occurred. Firstly, camera lenses were reducing in size but increasing in quality. Secondly, we progressed from VHS cassette onto solid state memory devices – or memory sticks. And lastly, the entire video industry was dramatically increasing the quality of recordings. Combine small lenses, compact storage and high-quality images and you have exactly what is needed in a dashcam.

The industry still didn't take off though – until along came Russia.

Insurance fraud involving vehicles and police corruption were widespread in Russia in the early part of this century – to the point where the Interior Ministry permitted citizens to install in-car cameras in 2009. Within three years, over one million vehicles in Russia were fitted with dashcams and it is estimated that close to twenty million dashcams have been sold in Russia in total.

Once 'Funniest Home Videos' started seeing so much dashcam footage that dedicated dashcam shows started, it didn't take long for the rest of the world to catch on.

The worldwide market for dashcams this year will be worth over $4.5 billion with almost 40 million units expected to be shipped. That is a lot of footage for the next episode of 'World's Worst Drivers.' Some cars are even now shipping with cameras built-in. My car has 8 external cameras and three of those store footage for later viewing.

Why do people install dashcams? They may do it to protect themselves and have a reliable witness if they are involved in an accident and use the footage to prove their innocence. Parents or employers may wish to use the footage to check on their children or employees – whether involved in an incident or just for general safe driving.

Back to my first point. When people are driving their car, they generally figure that they are behaving as if no one is watching them. Early data on dashcams has shown that a dashcam will reduce risky behaviour by teenagers by as much as 70 per cent. I am sure that data applies to a lesser extent to the rest of the population. That sounds like a great argument for installing a dashcam but if you are involved in an incident and you use your dashcam to defend yourself, will you end up being held responsible because you were speeding when the other person went through a Give Way sign?

Who knows – one day the Police may just pull you over and ask for your dashcam so it can be analysed for illegal behaviour!

Let me know if you use a dashcam at ask@techtalk.digital.

LOCKED OUT

I recently wrote a poem called 'Tomorrow's Smartphone User' and it focused on the fact that a modern smartphone has so many wonderful features that it may be easy for a smartphone user of the future to forget that it is still capable of actually making voice phone calls.

In the 18-29 age bracket, 52 per cent of smartphone users make a phone call less often than weekly. That same age bracket has average usage of over four hours per day. There is no doubt that younger generations use smartphones – just not for phone calls!

As our smartphones allow us to do more and more, I start to wonder when the lack of a smartphone means you are somewhat excluded from society.

There was a recent case in a New York City apartment block where the landlord installed a smart lock on the lobby door and changed the elevator to be controlled by smartphone. As often happens in America, when people are unhappy, they sue. Five tenants banded together and sued the landlord on two issues. Firstly, a 93-year-old tenant did not own a smartphone and was therefore unable to use the elevator or access the building without the help of another tenant. Secondly, the tenants were concerned about the privacy issues. Controlling the locks electronically would allow an interested party to possibly 'surveil, track and intimidate' tenants.

In the end the tenants won the case and a physical key was officially deemed a 'required service.' Some may argue this was a loss for technological progress.

Stop for a moment and think about the number of ways you can use a smartphone in your everyday life.

I no longer carry a car key. I carry a smartphone with an app that allows me to enter and start my car. Credit card? Nope. I use my phone or watch to tap and pay for items. I am not required to carry my physical Driver Licence. I haven't punched in a security code on an alarm system for many years. Access to security doors and cabinets in my business is allowed with a simple wave of my phone. When I listen to music, my phone or watch is providing the link to my listening device. Control my drone? Smartphone. Start my vacuum cleaner or pool filter or open my gates or garage or board a plane? Airbnb? You guessed it. Smartphone.

Back to my original concern. The lack of a smartphone can put a person in an interesting position. Smartphone ownership has not yet become as pervasive as ATM usage in our society. In the US, for example, 95 per cent of all adults own a mobile but only 77 per cent of adults own a smartphone. That means that there are still 23 per cent of people excluded from activities requiring a smartphone. When you break down the ages it tells a clearer picture. In the 18-29 age bracket, mobile ownership is close enough to 100 per cent with 94 per cent owning a smartphone. In the 65+ age bracket, mobile ownership is quite strong still at 85 per cent but only 46 per cent of people in this age bracket own a smartphone.

If we think back to the example in New York City, the lack of penetration in the 65+ age bracket immediately removes a segment of society from living in this particular apartment block. As much as I am a fan of technology, for a time we will still be using dual methods for a range of items in our society.

Tell me what you use your smartphone for at ask@techtalk.digital.

HEADS DOWN

I have seen a total of approximately one minute of comedian Rick Mercer's TV show and it was thirteen years ago – but that video clip has stuck with me since then.

Wind the clock back to 2006. The world was yet to see the iPhone. 4G was still a couple of years away and some locations still didn't have 3G. Users did not complain about slow download speeds though because the price of data was so incredibly high that you could go broke downloading a photo!

And the king of the modern hip and trendy professional was the Blackberry. Imagine a device that allowed you to send and receive e-mail, browse the Web, send texts with a full qwerty keyboard but still small enough to lose in your handbag. Better still they had an incredibly sophisticated worldwide server arrangement that compressed data and allowed a fixed monthly fee for unlimited Blackberry data.

People were so committed to their Blackberry devices (me included) that comedians would parody Blackberry users.

Enter Rick Mercer. In 2006 he produced a one-minute parody advertisement for a device he called the 'Blackberry Helmet' made of 'reinforced polymer to protect the skull of the mobile professional on the go.' My favourite part was the 'camera that broadcasts a picture of what's in front of you to your Blackberry so you can always be looking at your Blackberry.'

All this time I just thought that Rick Mercer was a comedian but it turns out he is a visionary and a safety activist.

Fast forward to today and pedestrian injuries are skyrocketing – mainly due to a phenomenon known as 'distracted walking.' In data from the US, pedestrian deaths have risen 41 per cent since 2008 with 6,227 deaths last year. That is a staggering 16 per cent of all traffic fatalities. Not all of these are due to users having their heads buried in a mobile, but the increase has been largely blamed on distracted walkers. So much so that various lawmakers in the US have started to introduce fines. Obviously the risk of death is not a large enough deterrent so the legislators have had to look to something more significant than death – money!

Fort Lee has banned texting while walking - $85 ticket. Honolulu will only hit you with a $15 fine for looking at your phone while crossing a street but the fine jumps to $99 for multiple violations. Stamford will fine you not for just texting and walking but also listening to music using headphones while crossing the street. The city that never sleeps may become the city that never texts with New York set to impose a $25 fine for first-time offenders of crossing the street while using their phone but, as with Honolulu, the fine jumps significantly if you don't learn from being booked once. A repeat offence will cost you $250.

In London it seemed a little crass to go around fining people for using their phones so they took a more polite option – by padding the lampposts to reduce the injuries for people who inevitably had their head down tending to something more important than looking where they were headed.

Back to comedian, no, visionary Rick Mercer. Maybe a protective helmet with a built-in camera and some added sensors isn't such a bad idea after all. Put a patent on it Rick. The perfect present for the millennial that has everything – a proactive device to try and avoid the accident combined with protection to reduce injuries when the accident inevitably occurs.

Tell me if you text and walk at ask@techtalk.digital.

The speed of G

On 9 April 2013, then Federal opposition leader and eventual Prime Minister, Tony Abbott, made a broad statement referencing the NBN. "[We] are absolutely confident that 25 megs is going to be enough, more than enough, for the average household." Further, "Do we really want to invest $50 billion of hard-earned taxpayers' money in what is essentially a video entertainment system?"

We know that our NBN is not keeping pace (early pun this week) with the rest of the world having dropped to number 62 but our mobile networks are in the top five. Mobile networks around the world are ramping up their plans for 5G with Telstra launching their $8 billion 5G network in Australia at the end of May making Australia just the third country to have 5G.

What is 5G and why is there so much hype?

The 5G name is the simple part. It stands for fifth generation. 4G launched in Australia in 2011. 3G started in 2003. 2G goes back as far as 1993.

The name is the easy part. The technology is a little more complicated. It all comes down to the frequencies used. 4G networks use frequencies below 6GHz and in Australia 4G ranges from 700MHz up to 2.6GHz. 5G uses extremely high frequencies – we are talking 30GHz up to as high as 300GHz. These high frequencies deliver one huge advantage and a small disadvantage.

Remembering back to your high school physics, the length of a wave is calculated as the speed of light divided by the frequency. Therefore, a higher frequency has a shorter wavelength. 300GHz translates to a wavelength of 1mm whereas

700MHz equates to 428mm. What does this mean? The higher frequencies and shorter wavelengths deliver higher bandwidth over shorter distances with the possibility of more interference from...just about everything. Buildings, cars, terrain and even water molecules.

The theoretical speed of 5G is 20Gbps (that is not a typo – that is Gigabits) compared to the best 4G can offer being 1Gbps. That is 800 times faster than Tony told us was the acceptable speed.

On top of the faster speeds, 5G will enable more simultaneous connections. Fantastic when you are at a concert along with tens of thousands of others trying to upload photos to social media, but also required as the number of Internet of Things (IoT) devices grows. The estimation is that we will have 22 billion connected devices by 2025. The 4G network would simply not cope with this number. 5G can support up to one million devices per square kilometre compared to 60,000 in the same area with 4G.

The limitation of distance is not a trivial one – but that will be solved by more antennas on towers and increasingly on buildings. These antenna arrays are much 'smarter' than 4G antennas. Whereas a 4G cell will just broadcast the signal at a constant power in all directions, the 5G network can better understand the data required and in what direction it is needed so it can vary the power accordingly.

In summary, 5G will allow more connections at faster speeds with lower latency – but it will need more antennas to deliver the same coverage.

Is this the death of the NBN? Not quite – but the number of NBN enabled households that will choose mobile will double from 15 per cent to 30 per cent. A moot point at the moment as the network has only just launched in ten cities and there are limited devices available, but keep a firm eye on this space.

Tell me if you are urgently awaiting the arrival of 5G at ask@techtalk.digital.

LISTEN OUT FOR THE NEXT REVOLUTION

iTunes is dead! Long live iTunes. Apple recently made the announcement that they are killing off iTunes. Oh no! What will happen to all my downloaded music? What about the songs I copied from CDs onto iTunes? My playlists!!!

Firstly, we should look at the history of iTunes.

Many people think that the iPod changed the way we listened to music forever.

Well, that is only partly true. The iPod is just a nicely packaged MP3 player. The first iPod was released on 23 October 2001. It had a mechanical scroll wheel and a massive 5GB of storage. Three years previous to the iPod launch was the first MP3. The MPMan F10 launched in March 1998 and contained 32MB of storage.

The secret to the success of the iPod was not the device. It was iTunes. On 9 January 2001, Apple launched iTunes. It was an Apple makeover of a program called SoundJam that Apple purchased in 1999. The coup de grâce for iTunes occurred on 28 April 2003. Apple launched the iTunes Music Store. To quote the original media release, it was "a revolutionary online music store that lets customers quickly find, purchase and download the music they want for just 99 cents per song, without subscription fees." We will come back to that comment about subscription fees. When it opened, the iTunes Music Store was the only legal digital catalogue of music to offer songs from all five major record labels. Apple boasted at the time that an incredible 200,000 songs were available in the store – which was significant then but is laughable when you consider that there are now over 30 million songs available.

To say it was an instant success is like saying that Roger Federer hits a tennis ball OK. Within 15 months, 100 million songs had been sold. The first billion songs took less than 3 years from launch date and ten years after launch, 25 billion songs had been sold. What a fantastic platform for Apple and a huge part of their financial success.

Surprisingly for Apple though, they missed a market trend. As mobile reception improved across the world, the concept of streaming music rather than buying individual songs started to gain momentum. Why buy one song – even when it is only 99 cents – when you can pay a small monthly fee and have access to listen to any of the 30 million songs available on a platform.

Spotify is the best known of the challengers. It launched on 7 October 2008 – possibly before its time – but as reception has improved and data plans have delivered more data, Spotify has grown to now be at 207 million users worldwide. Apple was late to the party and their streaming service only has 50 million users. Apple originally boasted that you could buy their songs without subscription fees but as it turns out, this is actually what we want.

With all the past success of iTunes, what prompted the decision by Apple to kill off a money-spinner for them?

The market has changed. iTunes hasn't. It is still the same basic program it was in 2001 but now has apps and TVs and movies and includes the kitchen sink. One program trying to do everything. Instead, they will shut down iTunes and repackage three different dedicated apps. Apple Music, Apple Podcasts, and Apple TV. Apple Music will focus on Apple's streaming service to try and catch up with Spotify but they will still allow you to keep all of your old purchased songs – phew!

Tell me how you listen to your music at ask@techtalk.digital.

EATs

Look! Up in the sky! It's a bird! It's a plane! It's...my taxi! We haven't noticed any people from Krypton flying around our skies but we are certainly closer to the world depicted in The Jetsons, the animated sitcom that originally aired in 1962.

This cartoon was remarkably good at predicting the world of the future. We now communicate via video calls; we watch flat screen televisions; electronic sliding doors are commonplace; we have vending machines for food; electronic toothbrushes; moving walkways; robot cleaning devices and life in general is assisted by various labour-saving devices. And just like our society, everyone on The Jetsons complains of the exhausting hard labour living with the remaining inconveniences. Items that today we might call first-world problems.

Despite this progress, we are yet to match the way George travels to work. He commutes in his aerocar with a transparent bubble.

But we are making progress.

The technology in drones has been progressing at a rapid rate. There are 26 countries that are trialling or planning drone delivery operations today. 17 of those are already delivering parcels as we speak and 13 are using drones for medical deliveries. There is huge variety in what is being delivered by drone.

New Zealand had the first pizza delivery by drone (I am not sure if a pizza is counted as a parcel or, to many, it may be medicine.) In Germany, industrial parts and pizzas are being delivered by drones – presumably by separate ones. Coffee is being delivered by drone in the only place where excess is the norm – Dubai. In the UK you can have your new phone delivered by drone.

Many people speak of regional drone deliveries but they will initially be used in highly congested areas. Drones have a complex mathematical formula of weight and range and charge. Drones need to be as light as possible because energy is expended in just keeping the drone in the air. To fly further distances, a drone relies on the amount of energy available in a battery. That seems easy – increase the size of the battery. When you increase the size of the battery, you increase the weight so you lose some of the gain. Increase the battery some more? Sure, then you need larger motors to lift the extra weight which, you guessed it, increases the weight and you need more battery and...you get the picture.

The real sweet spot here is a drone that travels in a city like Beijing or Sydney where traffic is a nightmare. Sending a delivery drone to a regional location where there is good road infrastructure doesn't make a lot of sense but if a location is remote and roads are abysmal, send in the drones.

The next step is transporting humans. It won't be long before you hail your taxi and a drone lands on a skypad near you. And Australia will be leading the way.

Dallas, LA and Melbourne will start a trial next year of an app-hailed Uber service. Forget new motorways or tunnels – this has the possibility of revolutionising transport. A one-hour trip from Melbourne's CBD to the airport will take just ten minutes by air. The initial trials will be with piloted drone like devices. The long-term view is that you will hail a driverless drone and then up to four passengers will board an electric air taxi (they need to come up with a better acronym than EAT) and arrive safely at their destination.

George Jetson – move over. Here we come!

Tell me if you would feel comfortable boarding a driverless drone at ask@techtalk.digital.

FEATHERS OR LEAD

I am sure you remember the old trick question that your science teacher gave you at some stage during your high school education. Which has more mass - a kilogram of feathers or a kilogram of lead?

Along those same lines, which is of more value – a GB of data in Australia or a GB of data in Bali? Using the same logic as the science question, the answer should be that a GB of data is the same anywhere.

Tell that to a family who recently returned from Bali to be met with a $30,000 phone bill. The culprit? Not their phones. Their innocent little iPad was used to watch a couple of movies and play a few games.

With 10.5 million Australians holidaying overseas last year, this is an ongoing issue. I do remember travelling overseas many years ago and I had to hire three different phones while I was in the US to cover the different networks they had at the time. Making a phone call on an Australian phone while in the US was either not possible or only possible if you were happy to mortgage your house to the telco.

As universal standards progressed and telcos worked better together, making a call while travelling became more reasonable – but who wants to make a call anymore? It is now all about the data.

In its infancy, international data roaming rates were horrendous circa $15,000 per GB. The carriers would dress it up to make it sound cheaper – for example that price per GB might be advertised as 1.5c per kB. Don't be fooled!

Luckily we have moved forward again with data charges – but it is complicated.

Talk to your preferred carrier a few weeks before you travel and know which countries you will be visiting. Some carriers have plans that include unlimited overseas calls and reasonable data limits while overseas.

These plans are great for frequent overseas travellers. For occasional travellers, there are now add-on packages that allow unlimited calls and texts for $5 or $10 per day.

Be wary of data still. These packages typically only allow 0.2GB of data per day.

If you do go over, the pricing is a little more reasonable but at $20 per GB for excess, it can still add up. Keep a few items in mind though. These plans apply in many countries – but not all. In most cases you are only charged the $5 or $10 if you use the phone within a set 24-hour period – but the timing is typically in Australian time not the local time where you are. The other trick that is commonly missed is roaming on a cruise ship. In that case, all bets are off! The cruise ship owns the network and they charge whatever they please.

The best advice when you are on a cruise ship is to enjoy the cruise! In general, as you travel the world, data is where you will be caught out so try and leave your data off and use Wi-Fi where available. When I travel overseas with my family, when we are choosing a café, my wife looks at menus and I look for Wi-Fi signs!

If you really can't be without your data, use a standalone local mobile data device so you can then use local data by connecting your phone to it. I often carry one in my pocket overseas – convenient for me to use and I always know the kids are going to be nearby – about as far as the Wi-Fi signal will travel!

Tell me about your international communications at ask@techtalk.digital.

Y2k Bug? Y2Worry?

Take a moment to cast your mind back to the late nineties. We were obsessed with news about Bill Clinton and impeachment. We were horrified by the Columbine High School massacre. We listened to Elton John and Britney and a new generation became fans of the Star Wars franchise with Episode I 'The Phantom Menace' while the Euro changed the financial landscape across Europe.

In Australia, we debated endlessly about a lovely lady with a rather strange taste in hats being in charge of this nation. We were urged to vote Yes in that referendum by the Chair of the Australian Republican Movement who many of us suspected wanted to run the country.

And...Y2k bug fever gripped the world.

First some context. When computers the size of a room were being introduced in the fifties, storage was at a premium. The IBM 350, introduced in September 1956, had the world's first hard disk drive. At 971kg it could only store 3.75MB and would set you back US$330,000 (current equivalent).

In writing software, the cleverest programmers deduced that two bytes of data could be saved when writing the date.

For example, the date man landed on the moon could be written as 16071969 or 160769 with the latter more space efficient. To any human or computer, it was obvious they were both the same date and at up to US$90 per kilobyte (in current money), programmers were encouraged to be more efficient.

The Y2k problem was highlighted in the mid-eighties but the deadline was still some time away. Other problems seemed more important. By the mid-nineties, there was finally some acknowledgement that we should do something about the issue. Despite many computer scientists around the world running simulations and analysing millions of lines of code, the Y2k issue caused heated debates at coffee shops by people with minimal research on the issue.

Just for a moment ignore all the logical data from specialists and pretend that there were two possible outcomes. It was either a hoax or the greatest self-inflicted calamity in mankind's industrialised history. Then there were two possible actions. Throw resources at the problem to negate the possibility of this being the worst party to welcome the New Year ever – or ignore it.

That gives us four scenarios.

First outcome. Do nothing – and it was a hoax.

News reports: "Lucky we didn't waste any time worrying about that problem."

Second outcome. Do something – and it was a hoax.

News reports: "Much ado about nothing. We need a Royal Commission to find out who perpetrated this hoax."

Third outcome. Do nothing – and it was real.

News reports: There won't be any. A complete breakdown of modern society would ensure the least of our problems was the fact that we couldn't turn on TV and watch the news.

Fourth outcome. Do something – and it was real.

News reports: "Welcome to the New Year." You can hear the artist formerly known as Prince singing in the background: "So tonight I'm gonna party like it's nineteen ninety-nine."

Of the four available options, the world took the only sensible course of action available to it. If it had been a hoax – mind you the best orchestrated and well-organised hoax in the history of mankind by a factor of a bazillion – then we would have wasted some effort and money on tidying up some lines of code. If we did nothing and the world collapsed, with the knowledge we had, how could future generations ever forgive us for what we had allowed to happen knowing there was a solution.

Many Y2k deniers pointed to a functioning society after midnight and told us we all fell for a hoax. That ignores one little number. 500. That was the minimum estimation, in billions of US dollars converted to current terms, spent across the globe to ensure we were ready for the biggest party in the world.

And we still didn't get it quite right.

Despite US$500 billion, we still had date-related problems after 1 January 2000. Credit card failures; corrupted satellite data; school heating failures; nuclear reactor false alarms; rejection of food shipments; incorrect age-based screening tests for pregnant women and even a video rental late fee of over US$90K!

I can't help but think of the parallels between now and the late nineties.

Today we hear constant talk of impeachment of a US President and we witnessed yet another mass school shooting at Parkland, Florida. We are still listening to Elton John and Britney Spears. Supposedly the last Star Wars movie hit our screens with Episode IX 'The Rise of Skywalker' while the European financial landscape is being challenged with Brexit.

In Australia, Malcolm Turnbull did eventually get his chance to run the nation – just not as a republic.

And looming over everything we do – just as with the Y2k bug – is Climate Change.

As with the Y2k bug, I see four possible combination outcomes based on two variables. Climate Change might be real or it might be a hoax. Like the Y2k bug, the scientists and experts have provided the proof and simulations to show the harsh reality of the situation but just for the sake of the argument, pretend it is a possibility that Climate Change is a hoax.

We then have the option for the world to take action...or not.

Option 1. Do nothing. Climate Change is a hoax.

Result? None of the dire consequences of Climate Change come to fruition and we continue to live our lives as per normal. Sure, we will eventually run out of coal and oil but that is for another generation to solve.

Option 2. Do something. Climate Change is a hoax.

Result? We reduced pollution across the world while we spent unnecessary money on creating other ways to produce power and propel cars. Some wasted money but the world seems like a nicer place to live.

Option 3. Do nothing. Climate Change is real.

Result? We destroy the planet and kill most of the people living on it. Full stop.

Option 4. Do something. Climate Change is real.

Thank goodness. The world was on the brink but mankind saved themselves from themselves.

When you break down the options, is it really worth the risk? In just the same way as the world could not afford the risk with the Y2k bug, it simply can't afford to risk taking a chance on not acting on Climate Change.

The biggest difference between Climate Change and the Y2k bug is that there was a definite deadline that everyone identified with the Y2k bug. 1 January 2000.

With Climate Change, we may well have already gone past our irreversible deadline.

Mathew Dickerson

Mathew Dickerson is not a Climate Change Scientist. Mathew takes his car to a mechanic, visits his doctor when sick and uses an accountant to lodge his tax returns. He also has confidence in the scientific process. He is applying logic to a problem that is discussed in society every single day.

WHEN IS INTELLIGENCE ARTIFICIAL?

I remember growing up playing the board game Monopoly. It was as much a psychological experiment with family and friends as it was a game. The analysis started when the tokens were chosen. Battleship – cutthroat. Boot – grudge. Car – outgoing and passionate. Hat – introverted and strategic. The Iron – toilers. Dog – trustworthy. Thimble – practical and sensitive. Wheelbarrow – tough.

The next step was the selection of the banker – the controlling personalities are the first to put their hand up. Once the game started, the real character traits quickly showed through. Who showed compassion and who went for the jugular? And, most importantly in terms of character assessment, who cheated! Hasbro, having produced the game since 1933, estimate that almost half of the players attempt to cheat during a Monopoly game.

Last year it prompted Hasbro to release a cheaters version of the game where cheating is actually encouraged.

In the latest move from Hasbro, we are witness to a small insight into the power of Artificial Intelligence (AI). It may, at first, seem like a trivial use of sophisticated computing power but it is but a small indicator of where society is headed.

In the latest iteration of Monopoly, a voice activated top hat sits in the middle of the board. This electronic banker controls all transactions through the game such as buying and selling of properties and tracking of bank balances. No actual money changes hands – which means no player can subtly slide some cash from your pile across to their pile! Playing Monopoly with AI sounds like an interesting development, but what else is happening in the world of AI? As much as many people are fearful of computers taking over the world, most of us are interacting with AI every single day.

When you type an address into Google Maps, an incredible number of calculations are performed relating to road conditions and time of day and algorithms are then utilised to give you the optimal route and an unbelievably accurate travel time – helped by mobile phone metadata. None of that would be possible in a timely fashion without AI.

In the world of finance, accuracy and efficiency is critical – along with real-time reporting. AI is used to predict stocks for various portfolios by scanning millions of key data points. Your Superannuation is most likely being helped by AI.

In what many people find just a little creepy, when they use their social media channels or perform online searches, AI builds up more and more information on you so that the ads and feeds that show up are tailored to suit you. All of this information is being curated by AI and once again, the time it would take a human to analyse this volume of information would make it impossible to perform in a timely fashion.

The list goes on. Siri; smart home devices; autonomous cars; online shopping experiences and related customer service; the healthcare industry and so much more. Without the power of AI, companies that we accept as commonplace today would not even exist. Ridesharing applications like Uber and Lyft need to calculate the price of your ride and the shortest possible wait time. All performed with AI. Our mobile phone voice to text services, albeit with sometimes amusing results, uses an incredible amount of computing power and AI is at the core. Even hit songs, such as "Not Easy", are being influenced by AI although not yet fully written and performed by computers. Don't count it out though.

Tell me if you would like your own 'J.A.R.V.I.S.' from the Iron Man movies controlling your home at ask@techtalk.digital.

FREE GROCERIES ANYONE?

Woohoo! I just won my grocery supply for the next year! With four hungry kids our grocery bill is a significant expense in our household. I can hardly wait to see those free groceries appearing at my doorstep.

I've been waiting for a week now but still nothing. I have given the nice people at the major supermarket chain all the detailed information they asked for. I even paid them that small amount of money they requested to have the legal contracts for my prize sent to me by courier instead of via the normal post so I could start receiving my groceries immediately. The amount was nothing though compared to how much I am going to save over the next year. We will finally have that family holiday to some great spots in Australia that I have always dreamed about.

It has been two weeks now and I have tried ringing the supermarket chain but I must have written the number down wrong because the message says the number is disconnected. To make it more frustrating, the staff are not very good because when I rang the main switch number, none of the staff seemed to know what I was talking about! I was having a clumsy day because I also wrote down the e-mail address wrong and I can't remember how I found the site.

I know it would seem fairly obvious to my readers that this is a scam – yet we have people every day, right here in Australia, falling victim to exactly this type of scam.

Why?

Two reasons. Firstly, the idea of an unbelievable prize is so alluring that it often masks common sense. The excitement of a large windfall or a free holiday is the focus for the victim and they definitely don't want to miss out. Secondly, the scammers are very good at what they do. If only they turned their hand to more honourable pursuits…

The scams that work best have a great amount of attention paid to the details and the scammers have a good understanding of psychology. It is easy to setup a social media or Web presence but the 'best' scammers have accurate logos and contact information. You are hard pressed to spot anything out of place. By using legitimate contact details on a fake site, it gives it an air of legitimacy.

How do you navigate through this minefield to find if you are dealing with someone legitimate? Firstly, whatever you do, sprinkle it with common sense. Actually, douse it in common sense. Does it sound right? Is it too good to be true? Does something not look right? Once you apply that aspect, then look for a few tell-tale signs. A file that ends in .exe or .zip that you are being asked to 'open' is a red flag. Requests for personal information when you haven't initiated the contact is a no-no. Hovering over a link to see if it is a legitimate domain name before you click is also a good idea. Oh, and did I mention to use your common sense? Definitely do that.

> My dream is that we are vigilant in our interactions with scammers to the point that they give up and use their obvious talents to solve world peace or fix climate change.

Tell me if you have been scammed by sending an e-mail to ask@techtalk.digital. The first one hundred e-mails I receive will win a six month round the world holiday for themselves and fifty of their closest friends…just kidding of course. I wanted to see if you were paying attention!

CHARGING AHEAD WITH SOLAR

Toyota has just started testing a new Prius with a solar roof that will be used to charge a plug-in battery. With the increased efficiency of solar panels, it makes sense. To give you some perspective, the amount of power that hits the earth's surface at midday at the equator is approximately 1kW per square metre.

The solar cells on the Prius, being only 0.03mm thick, will deliver about 860W of power. This is enough to add 45km of range in an average day of sitting in sunlight.

The World Solar Challenge biennial car race relying solely on sun powered vehicles has been raced between Darwin and Adelaide since 1987. The winning entry in 1987 averaged 66.9km/h for the 3,000km while the latest race had an average speed of 81.2km/h. Clearly we are progressing in producing power from the sun.

Solar panels seem to get a bad rap because the sun doesn't hit your spot of the earth for twenty-four hours of the day so people are rightly concerned about how they might turn their lights on at night, but there are some major advances occurring that will start to deliver superior solutions.

Before we go there, we should go back in history. In 1891, Clarence Kemp in California patented a small glass-topped box with water running through it to heat water. Using the sun to save on regular heating of water. Not exactly creating electricity from the sun but using the sun to save the need for other forms of energy. Similarly in 1953 when SW Hart and Co. created Solahart hot water systems. Solahart now has over one million hot water systems installed in 70 countries around the world.

It was three years later that the first solar cells to produce electricity were available commercially. As with all new technology it was expensive. US$2800 (converted to 2019) for a one-watt solar cell. By comparison, that cost is now down to US$3.18 per watt. Despite the price, it allowed the start of the solar revolution.

Calculators and novelty items started appearing with tiny solar cells but by the late fifties, satellites were being powered by solar cells.

From the seventies, when a large drop in prices occurred, we started seeing solar panels in a variety of situations. Most off-shore oil rigs use solar panels to power lighting on the rigs. There is a certain amount of irony here! We also started seeing solar panels at railroad crossings, communication towers and water pumps. You often see a farm with a dilapidated traditional windmill no longer in operation and sitting proudly beside the windmill is a small solar cell and an electric pump. Less maintenance and more reliability!

To give you an idea of how quickly technology is progressing, a solar farm in Nyngan in regional NSW was started in January 2014. The 250 hectares is home to a 102MW solar plant. Fast forward almost five years and go up the road to Nevertire and in less space (180 hectares) a new solar farm is being built that will generate more power (132MW).

The initial concerns about night time power haven't disappeared but with pumped hydro and battery technology, among other systems, giving access to power through the night, expect to see more solutions involving solar panels.

In September 2009 I wrote an article where I had calculated the surface area of solar panels needed to power Australia was 32km square. I can only imagine the smaller size we would need with our more efficient modern solar panels.

Tell me the unusual places you have seen solar panels at ask@techtalk.digital.

The world needs more lerts

Great! There is a new app I want to use that all my friends are talking about. I can hardly wait to download it and start using it. Terms and conditions, I hear you say? Ha! Who has time to read all that legal gobbledygook? I am sure it will be OK and I won't be able to make much sense of it anyway. I want to just download the app and start using it. Click Allow and Next a few times...

Great. I am in. Wow this is a great app. I can see why so many people are using it.

Fast forward an undetermined amount of time...

What? Why was my phone bill so high? What are these charges on my credit card? Why is my face being used to advertise sex toys?

We live in a global marketplace. The current world population is 7.7 billion. There are some of those people that are dishonest or have ulterior motives...

One of the latest app trends is FaceApp. People are uploading photos of themselves using the app and applying a filter to show what they will look like in their elder years. For a start, I don't quite understand the concept. The global cosmetic products market is valued at US$620 billion annually. With all that money being spent on looking better, why does an app that makes you look older suddenly go viral? I won't try and understand humans – I am here to talk about tech.

When you apply the age filter, your photo is uploaded to a server which is then used to render a new image to show your aged self. This is the power of sophisticated AI for outright fun. I have noticed many images lately from friends and celebrities and politicians but I would guarantee very few have read the Terms and Conditions.

Just by downloading the app, you are agreeing to some items that would make you squirm. To quote, "You grant FaceApp a perpetual, irrevocable, nonexclusive, royalty-free, worldwide, fully-paid, transferable sub-licensable license to use, reproduce, modify, adapt, publish, translate...display your User Content and any name, username or likeness...in all media formats and channels...without compensation to you...you agree that the User Content may be used for commercial purposes." This is just a tiny part of what you have agreed to – just by DOWNLOADING the app.

To break this down, if you use the app, you might be driving down a freeway in Vanuatu one day and see your photo on a billboard advertising a strip club in Port Vila or your face could be the image used by a worldwide organisation that is trying to rid the world of chlamydia. At a local level, if the Prime Minister used FaceApp, there would be nothing to stop the developers selling Anthony Albanese the image of Scott Morrison to be used to advertise the latest ALP policies! Imagine ScoMo trying to explain that he didn't really give permission to use his image but he actually did agree to it by downloading the app!

You might think this scenario a little far-fetched but time and time again I see users with ridiculous phone bills or credit card statements from users who clicked OK without reading information correctly.

Bottom line – keep using apps as there are some wonderful solutions to problems we didn't even know existed BUT be alert. The world needs more lerts! Jokes aside, try and be aware of what you are agreeing to when you use an app and then make an informed decision.

Tell me the worst T&C situation you have seen at ask@techtalk.digital.

SEE MY SONG

"If I have seen further it is by standing on the shoulders of Giants." It would be hard to imagine what Sir Isaac Newton would think, almost 350 years later, of how the world of technology has taken to heart this concept of building upon previous knowledge.

Today I am going to talk about one specific product which is a little unusual for my tech columns, but it is the best example I have seen for some time of that concept of a product that builds upon a number of different pieces of technology.

Step 1. We have this little thing called the Internet. More than anything else, the Internet gives us access to an incredible amount of information – and it is searchable in mere seconds. We have only had access to something resembling the current form of the Internet for less than thirty years.

Step 2. Liquid Crystal Display (LCD) screens were accidentally discovered in 1888 and were a scientific curiosity for decades. It wasn't until 1964 that the first working LCD was built by George H. Heilmeier and in 1988 Sharp developed an underwhelming 14-inch LCD TV. By 2008, annual sales of LCD televisions surpassed the sales of CRT units for the first time.

Step 3. The next step in the music revolution. After cassette tapes and vinyl then CDs and DVDs, music moved online. 1993 was the first time an online music archive was available to download and this led to Ritmoteca.com quickly followed by Napster and then, in 2001, Apple's iTunes Music Store. Downloads soared stratospherically. Within three years, one billion songs had been sold via iTunes.

Step 4. The next, next step in the music revolution. In 2008, Spotify launched. It may have been before its time but is reaping the rewards now. Why buy and download individual songs – the way that old-fashioned Apple model worked – when you can access just about every song ever produced for a reasonable monthly fee. 217 million users currently agree that Spotify is a sensible way to source their music – and that is just the largest of a number of online streaming music providers.

Step 5. Forget about physically visiting a library to access information. With the Internet, you can build a database and the world can access it. Databases of song lyrics started being created by the mid-nineties and it would be hard to find a song where the lyrics are not available online somewhere. As a kid, I used to be attracted to the bands that included an album sleeve with the lyrics to the songs printed on it but online lyrics make this so much easier.

Put all of these previous discoveries together and you end up with a device that is the perfect technology example of Newton's quote.

The Cotodama Lyric Speaker allows you to connect to your speaker via cable or Wi-Fi (no big deal there), access songs from just about any of the streaming providers (again pretty common) and when the music plays the speaker automatically sources the lyrics of the song and displays them on an LCD at the front of the speaker. Using technology developed by Japan's National Institute of Advanced Industrial Science and Technology, the words display differently depending on the mood of the song.

> Since Alexander Graham Bell patented the first loudspeaker in 1876, I can think of no better example of the most technologically advanced 'simple' device than the Lyric Speaker.

Sourcing songs online, sourcing lyrics online, connecting wirelessly and displaying the lyrics!

Tell me if you like to see the lyrics when you listen to music at ask@techtalk.digital.

If you can't lick 'em…

For over one hundred years, pragmatic advice freely handed out to people or organisations in a seemingly doomed situation has been "if you can't lick 'em, jine 'em."

I suspect that was the logic applied in the latest move just announced by Foxtel.

We should go back just a few steps. In 1995, a venture between News Corporation and Telstra created an organisation that would use a coaxial network to transmit a TV signal to consumers who would pay for this service. On 23 October 1995, the first Foxtel transmission started with a twenty-channel service. Foxtel was not the first pay TV operator in the nation though.

Galaxy began a two-channel broadcast on Australia Day in 1995 but only lasted three years. Optus Vision and Austar also beat Foxtel to the punch but the combined might of Twentieth Century Fox and Telstra soon made Foxtel the dominant player and the acquisition of Austar in 2012 effectively delivered total dominance in the pay TV space.

My initial excitement with pay TV was in relation to advertising.

When Foxtel launched, advertising during programs was banned under Australian Government legislation for the first two years. That has now changed, of course, but legislation still prevents a pay TV business from earning more than fifty per cent of their revenue from advertising.

Many people are frustrated watching ads on TV so the lure of no advertising was a great attraction initially for pay TV.

Many people would think that Foxtel was in a great place from 2012 onwards – and I agree. Complete dominance of a market with over five million subscribers sounds like a great place for a business to be in. As so often happens, though, complacency is a cruel leveller. When a brand achieves market dominance, they can become complacent and competition ends up coming from a product that is the same but completely different. Think of Nokia, who dominated the mobile phone world, only to have their world turned upside down by the Apple iPhone.

Just when Foxtel could relax and count their dollars rolling in, along came a disruptor. On 24 March 2015, Netflix was introduced in Australia. Within three years, Netflix had grown to nine million users and growth continues as they approach twelve million subscribers today. Meanwhile Foxtel numbers are slipping and the subscriber base is now below five million.

In this situation, what would you do if you were in charge of Foxtel? They provide a different service to Netflix with live sport and current news programming which adds significantly to the cost base which is passed onto consumers but ultimately everyone is competing for consumer eyeballs.

Foxtel decided they couldn't beat Netflix so they may as well join forces. In the biggest move in Australian pay TV services since Foxtel bought Austar, Foxtel now provide all Netflix programming as part of the Foxtel offer. One remote, one service and all channels. Combined with the combination is a relaunch of the entire menu system. No longer do people access 'channels' but they access content.

Searching for preferred content on one remote across everything that Foxtel and Netflix delivers is an offer that Foxtel management hope will keep you paying your Foxtel subscription. Add more on-demand programs with an ever-growing library and maybe the combination of Netflix into the Foxtel offer will be the move that saves Foxtel – and that is a big statement compared to how the management would have felt about their future in 2012 when they bought Austar.

How quickly the world can change.

Tell me what pay TV services you use at ask@techtalk.digital.

THEORETICAL OR APPLIED TECH

In technology and innovation, it can be as much about when as it is about what. Six years ago, I was involved with a charity event and a standard recipe for such an event was for an auctioneer to take the microphone and spend thirty minutes trying to extract money from wallets in the room.

For people paying good money to attend an event, it seemed an unnecessary interruption to the night for the sake of fundraising. I had an idea to use technology instead.

So I cobbled together a few basic online auction tools and displayed products online and...it was a flop.

With only 45 per cent of the adult population owning a smartphone at the time, not enough people in the audience had smartphones and not enough people were comfortable putting their credit card details online.

I shelved the idea - until a month ago. Six years after my first attempt, I tried it again. This time around it was an unmitigated success. The only difference? Timing. With over 90 per cent of the adult population now owning a smartphone and online shopping being tried by 96 per cent of the adult population, the timing was right.

It started me thinking about other flops that were based on timing.

There are over 20 per cent of adults who currently own a Tablet and most of those owners would tell you the market segment was started by Apple with the iPad announcement in 2010. The Microsoft Tablet PC was introduced ten years earlier and despite featuring several innovations we now take for granted, the technology was just not quite advanced enough for consumers who want it to work today.

When we think of electric cars, we immediately think of Tesla. The first deliveries of the Tesla Model S started in 2012. We need to go back a few years though to the GM EV1. It was launched in 1996 and was a huge hit with customers who managed to secure one. Reminiscent of Kodak's decision to patent the digital camera and then shelve production plans for fear of what it would do to film sales, GM was concerned what an electric car would do to the sales of spare parts and officially recalled all EV1 vehicles in 2003 and destroyed them.

Before the release of the iPhone in 2007, Apple had a previous attempt that was a complete flop. The Apple Newton (or MessagePad) was released in 1993 but, in 1998, Steve Jobs killed off the product upon his return to the company. The concept was good but the battery, screen resolution and touch screen were all below expected.

Many of us watch TV shows on our phones without a second thought. In 1982, the Sony Watchman was released as a portable TV on the back of the success of the Sony Walkman. With a 5-centimetre black and white screen and a limited selection of channels and poor reception, it was decades ahead of its time – and a flop!

In 2019 more than 50 million smartwatches will be sold, with the Apple Watch the dominant model. In what seems like a case of déjà vu, Microsoft had two failed attempts at a smartwatch before Apple released a successful model. In 1994, Microsoft co-developed the Timex Datalink and then in 2004 the Microsoft SPOT smartwatch was released. With poor graphics, connectivity and display, the technology could not deliver on the promise.

> With all of these examples, a thought leader needed technology to catch up with their vision.

Tell me your best example of technology before its time at ask@techtalk.digital.

EVOLUTION OR REVOLUTION

Paraphrasing the former US Secretary of Education, Richard Riley, we are currently preparing students for jobs that don't yet exist, using technologies that haven't been invented, in order to solve problems we don't even know are problems yet.

One of the greatest aspects of the launch of the iPhone on 29 June 2007 was that Apple was solving a problem that consumers across the world didn't know existed. Here we are, only twelve years later, and smartphone market penetration sits at ninety per cent. How did we ever survive without being able to use a device in our pocket to scan documents or view the night sky or order food or transport?

The constant challenge for smartphone manufacturers across the world is to improve upon the concept or develop the next killer feature.

This is a serious race.

The worldwide market for smartphone sales is estimated to be worth US$570 billion this year.

The latest flagship from Samsung has just been released. The Note range - the 10, 10+ and 5G. Are we now at the stage where we are seeing evolution or revolution? Small incremental changes or giant leaps?

For a start, the Note is now available in two sizes – the 6.3" Note 10 and the larger 6.8" Note 10+ (and 5G). When you consider that there are Tablets on the market with a screen less than 8", you can see just how large the phone has now become. The 6.8" screen has 3840 x 1440 pixels on display. For comparison, the first iPhone had a 3.5" display with 320 x 480 pixels.

Many consumers are interested in the camera (or more accurately cameras). The original iPhone had a 2MP (megapixel) camera and 4GB of memory. The Note 10+ has four cameras on the rear ranging from 12MP to 16MP and a 10MP camera on the front. Four cameras I hear you ask? 16MP ultra-wide; 12MP wide-angle; 12MP telephoto and DepthVision lens. Seems obvious I say with my tongue firmly in my cheek. When you consider it was 2002 when the Nokia 7650 was the first phone to feature a camera (at a massive 0.3MP), it seems incredible that we now have phones with five (or more) cameras. The Note 10+ puts the focus on video with a live bokeh effect and sophisticated video editing utilities built into the phone. And AR of course.

When it comes to sound, Samsung followed the Apple lead and removed the headphone jack with their research showing the majority of users are already using wireless audio. Samsung took an additional step though with Sound-on-Display technology. The entire screen now works as a front speaker which eliminates entirely the top speaker. On top of that, the fingerprint sensor to unlock the phone is beneath the display.

Apart from the evolutionary steps of a better S Pen with more features and the phone's increased memory and improved battery life and charging times, the last step is the blurring of PC and phone. The latest Samsung allows you to run a virtual machine in a window when you plug your phone into your PC and you can also mirror your phone's screen to your Windows PC. This may well be the first step in the elimination of a computer and people using their phone with a dock.

Although many of the features are incremental evolutionary changes, I still think we are seeing some revolution in this space but where will it all end?

Tell me the next killer feature you think we need in a phone (so I can quickly patent the idea) at ask@techtalk.digital.

ELECTRONIC VOTING - I WOULD VOTE FOR THAT

When the fictional character Tom Dobbs (played by Robin Williams) was elected as President of the United States in the 2006 movie, Man of the Year, it played on the fears of people across the world when the subject of electronic voting is raised. In the movie, the Delacroy voting company won the contract to provide voting machines for the Presidential election but a software glitch delivered the winner based on spelling rather than votes.

Across the world, only fourteen countries have dabbled in electronic voting and, of those, it is only widespread in four.

With the greatest respect to the various commentators and election analysts, one of the most attractive components of electronic voting is the instantaneous and incredibly accurate method of delivering a result. In recent Federal and State elections in this country, after watching hours of analysis by a variety of experts and witnessing numbers dribbling through into tally rooms, we all went to bed on the Saturday night not entirely sure who was going to be in charge. A further attraction is the dream that we won't waste tonnes of paper for candidates to print beautiful glossy brochures to hand out to voters as they make their way through the scrum of volunteers.

The positives are obvious for all to see so why is it that we are reluctant to go that way. Australia in particular, with residents known for being early adopters of technology, logically would embrace electronic voting.

The fear of security breaches seems to overcome the attraction of the advantages. There may be other reasons our governments don't head down the path with electronic voting but security is the main issue put forward.

The US has long had a variety of solutions to this problem and, via DARPA (Defense Advanced Research Projects Agency) they have recently awarded a $10 million project to an organisation to create an open-source voting machine that could prevent hackers from tampering with votes.

The first real test occurred at DEF CON 2019. DEF CON is the world's largest hacker convention with over thirty thousand people in attendance.

At the Voting Machine Hacking Village setup in DEF CON since 2017, hackers typically take a few minutes to expose vulnerabilities in voting machines. At the 2019 event, the latest hardware from DARPA was on display at DEF CON and, ignoring a few minor setup issues, no hacker was successful in breaching the voting machine.

With Trump's presidency still plagued by rumours of Russian interference, DARPA is keen to produce a system that will allow lawmakers to pass an election security bill in time for the 2020 presidential election.

I would also suggest there is a large potential return on investment if they manage to produce a secure system. Once security was removed as an issue, governments around the world would be interested in licensing the technology to use in their elections.

And maybe there is something there for a progressive Australian government (oxymoron of course) to look at.

It is common knowledge that the CSIRO has generated over half a billion dollars for Australians from its Wi-Fi patent. I would propose that we could task the CSIRO with creating a secure voting methodology, patent it and use it in elections in Australia. Licensing that technology to other governments around the world to use in their elections would surely generate significant revenue for our government?

Perhaps I am over-thinking the first step though. Much of the focus for electronic voting is on securing the process for remote voting. I would propose a staged process to introduce electronic voting.

Stage 1.

We continue to use voting booths and have people physically visit an election booth on election day with one significant twist. Instead of printing reams of paper and asking voters to write on those pieces of paper, I propose that we introduce electronic voting terminals into all polling booths. When you enter a polling booth in the normal way, you need to produce ID and you are electronically recorded as having voted. This immediately is more secure than the current method. Currently you don't need to show ID and you could enter every different polling booth during election day and your name would be ticked off in a book. After the event you may be contacted and told you voted more than once but with no ID check, where is the proof that you actually did vote more than once?

Once your name was marked off, you go to an electronic terminal to vote for the candidates. Being electronic, you would receive a warning if your vote was not completed correctly. If you are going to make the disappointing deliberate choice of voting informally, you can still choose to do so but accidental informal votes would be a thing of the past.

You finalise your vote on screen and 'submit' the vote. Once it is entered, two pieces of paper print out from your terminal. These two identical printouts have two pieces of information on them. Firstly, it shows how you voted. Secondly, it creates a unique random number on the voting slip. As you leave the polling booth, you drop one slip into the ballot box the same as you would do now and you put the other piece safely in your handbag or wallet.

At the end of the election day, all of the votes are already recorded and it would take mere seconds for all votes and preferences to be counted. The results would be known immediately after the election.

If there was any suspicion of a breach of the computer system, the physical pieces of paper would be on hand to allow a physical count to take place as well. Furthermore, a Web site would publish ALL votes on a Web site with the unique random number identifying how each vote was cast.

There is nothing to identify an individual by name with those details but any voter who was concerned about the legitimacy of the process could look up the information on the Web site and check their unique random number against the piece of paper they took with them from the voting booth. This would satisfy each individual that their vote was accounted for correctly.

At this stage the main advantage gained is that the votes are entered quickly with less chance of fraud and the count is instant. Keep in mind that with the current physical voting scenario, all of the information is entered (by humans) into computers and then they are instructed to perform the count.

Stage 2.

Once the voting public was familiar and comfortable with this method, the process could be extended to remote electronic voting. Using the same processes – ID check to access a portal then voting and printing a unique identifier etc. it would allow people to vote from the comfort of their homes – or from anywhere in the world, in fact.

We are now at the stage where we are comfortable with our entire life savings being accessible from our smartphone and our lives seem to be contained within that little device in our hands – yet we don't seem to be able to vote electronically.

I am sure we can overcome this problem. It just needs the political will to make it happen.

Tell me if you would like to vote electronically at ask@techtalk.digital.

TECHNOLOGY. GREEN OR MEAN?

I had the pleasure of being the MC at a tech conference recently. During one particular panel session, the discussion centred around the advantages of high-speed connectivity and what we are now doing differently because of this paradigm shift.

Not surprisingly the issue of Climate Change was raised and everyone felt quite good about themselves with the tech sector making a positive contribution with such items as video conferencing to reduce the amount of travel needed, particularly air travel, and reducing paper requirements by aiming for paper less (not paperless) offices. Even the way a technician now services a client was discussed with the likelihood that a problem is more likely to be solved from within the service office reducing truck rolls.

While we were all patting ourselves on the back and thinking about how we were helping solve the largest problem the planet will see this century, an audience member asked a question that was so good that I promised her I would dedicate a column to the concept.

With the explosion of social media sites and online tools, all of the servers required that live somewhere in the cloud must be consuming significant resources and producing an incredible amount of greenhouse gases.

She had an excellent point.

For a start, we should examine that explosion. The Internet in its current form came to Australia in 1989. YouTube had its first video upload in 2005. There are now more than 300 hours of video uploaded every minute. Facebook started in 2004. There are now 2.4 billion users. Instagram has over one billion active users. It didn't exist last decade. Twitter and LinkedIn started in 2006 and 2003 respectively. Between them they have almost 700 million users.

The world has changed dramatically in the last thirty years. All of those cat videos and pictures of bacon and eggs arranged 'just so' have to be stored somewhere. "In the cloud," I hear you say. Correct – but the cloud has a home. Data centres all over the world. The best estimate is that there are 80 million servers across the world powering the Internet. At a very rough estimation, that would equate to 600TWh per year of electricity consumption or almost three per cent of the total world consumption.

Wow!

Admittedly there is movement towards producing more and more electricity with renewable methods but still, WOW! To give you an idea, that is 100,000 wind turbines!

On the flip side is the reduction in the number of on-site servers across the world. Before our current connectivity options, a business would install an expensive and power-hungry server at each location and it would be utilised for a part of the day but for a large part of the day it sat idle. There is a transition away from those servers to servers that are hosted in data centres which are used more efficiently. Furthermore, apps and services are replacing dedicated services running on a physical server. Does it equate to a saving of 80 million servers? Not even close, but there is a small saving there. I haven't even started to calculate the resources utilised in building the physical equipment and mining the resources required.

There is an interesting research project for a PhD student here. All the savings in travel and paper and efficiency versus the total resources utilised and power required to run the connected world we live in.

From a pure environmental perspective, is IT Climate Change friendly or doing more damage than we wish to admit?

Tell me if you think we can claim IT as 'green' at ask@techtalk.digital.

2G OR NOT 2G?

2G, or not 2G: that is the question: Whether 'tis worthier to the mind or to suffer through 5G. Unlike poor Hamlet, I have not lost my mind but there is a lot of confusion with the various levels of 'G' that are quoted by telecommunication providers, and to confuse us all further, Huawei has announced they are researching 6G!

I thought it worthwhile to give a very brief overview of the various generations of mobile phone technology to better understand how 5G fits into our landscape.

Australia had a very early '007' mobile service that was launched in August 1981. It is often referred to as 0G as it was not a true cellular mobile system. It operated in the 500MHz band and was shut down by 1993.

Mobile phone technology came of age in February 1987 when the analogue cellular system (AMPS) was launched with handheld mobiles.

This system is typically referred to as 1G and was a true cellular system with frequency re-use. It operated in the 850MHz range. This network was closed in 2000.

By the time 2G arrived in 1993, less than 4 per cent of the Australian population had an analogue mobile. 2G increased security and, after the first SMS had been sent the previous year, introduced Australians to text messages. 2G was typically referred to as GSM but some countries (including Australia) also had CDMA to deliver additional coverage. CDMA lived a short life – 1999 to 2008. 2G typically

2G OR NOT 2G?

operated around the 900MHz range and later the 1,800MHz band. 2G also introduced data with GPRS and EDGE at theoretical maximum speeds of 0.384Mb/s. 2G was largely shut down in 2016.

3G was introduced in 2003 and saw the dawn of mobile broadband with significantly better data speeds, up to 7.2Mb/s.

3G operated in the 850MHz and 2100MHz ranges. My estimation is that 3G will be completely shut down in Australia by 2023.

4G kicked off in Australia in 2014. Why? Three reasons. Speed, speed and speed. With greater data demands from users of smartphones and other mobile devices, the network needed to deliver. 4G has theoretical maximum speeds up to 1,000Mb/s. To give you some idea of the workload placed on modern carriers, in one month Telstra alone connects 500 million calls, sends 50 million text messages and carries 50 petabytes of data! 4G operates across a number of frequencies from 700MHz through to 2,600MHz.

Which brings us to 5G. Recently introduced in Australia it is currently only available in ten cities. The demand for 5G is not only driven by the need for speed but the need to be connected to, well, everything. 5G will have a theoretical speed of 10,000Mb/s but, more importantly, it will have reduced latency (more than five times better than 4G) and will allow significantly more simultaneous connections.

The combination opens up a whole new world of Internet of Things (IoT) devices that will explode onto the market. 5G will initially operate in the 3,500MHz frequency range but in future years will operate up to 28,000MHz (28GHz). This is significantly higher than all of the other generations of mobile technology but radio waves are non-ionising waves and, forty-six years after Dr Martin Cooper made the first phone call from a mobile, we are still yet to see any negative health effects from radiation caused by mobiles.

My advice to people across Australia in relation to the latest mobile standards is the same as my advice to my children after an argument. Embrace!

Tell me if you are excited about the advent of the 5G network at ask@techtalk.digital.

HOW TO OUTRUN A LION

was browsing my social media accounts earlier in the week. I saw a post from a friend and something didn't seem quite right. People do put some crazy stuff on their social media accounts but this seemed particularly bizarre. In this case a friend of mine was boldly professing his love of the Manly Sea Eagles and claiming to be their number one fan.

I am the only person I know that is brave enough to admit they support Manly. I contacted my friend – who had no idea what I was talking about.

The penny dropped. His account had been compromised.

The chief of Twitter, Jack Dorsey, recently had his Twitter account hacked. Suddenly Jack was sending out bizarre and racist tweets that would rival Trump. You would think the head of one of the largest and oldest social media sites would have some pretty good security but hackers were still able to take over his account.

A massive twenty-two per cent of people who use social media have had an account hacked. Fourteen per cent more than once.

It is impossible to make yourself hack proof but, just like installing an alarm means that burglars may chase an easier target, there are some steps that make it harder to compromise your account.

Firstly, you are only as strong as your password. A password that is a word that exists in the dictionary is easy for a hacker to find with a dictionary attack. How do you make it harder? Use a combination of characters that do not make up words. A computer can still try all known permutations but it will slow it down a little. To slow them down a lot, use more characters in your password and use a larger set of characters.

Let me give you a few examples. If you use just numbers and you have only 4 characters, that equates to 10,000 permutations. That would take a computer the brief blink of an eye. Now, instead of just all numbers, use the combined sample space of all numbers; letters (upper and lower case) and 'other characters' (#, $, % etc.) then instead of 10 characters you have 95 to choose from. A 4-character password from 95 jumps to 81.4 million variations. That is better but not a big challenge.

> The important part is length. Go from 4 to 8 characters in your password and the permutations now jumps to a number with a 6 followed by 15 zeroes. Two years to crack. Length does matter!

Go from 8 to 12 and you now have so many permutations that the number starts with a 5 and is followed by 23 zeroes. Now we are at 171 million years to crack.

The first step, therefore, is to use a long password with characters taken from all the different possibilities. Then enable two-factor authentication. Ensure you have anti-virus software on your PC and use a password management program rather than have every site with the same password.

Remember it is not about being hack proof, just harder than the next person. Think of the joke about three friends on the African savannah who stumble across a lion and one of them fixes his shoes ready to start running. When another friend comments that not even Usain Bolt could outrun a lion's top-speed of 80km/h, the first friend says that he doesn't need to outrun the lion...he only needs to outrun his two friends!

Lastly, if something doesn't look right or promises something too good to be true, be sceptical. Tell me a hacking experience you have witnessed at ask@techtalk.digital.

Scooting ahead

What do the names Beam; Bird; Bolt; Circ; Dott; Jump; Lime; Poppi; Scoot; Skip and Trotti have in common? With less than seven years of history in this market, not many of these are household names…yet. This is despite the fact that some of these companies are worth over US$2 billion.

One of the aspects of technology that I have always enjoyed is that one plus one often equals much more than two. Many products are developed with a specific purpose in mind and, in seemingly the blink of an eye, the technology is being used in an entirely different way.

Think GPS (Global Positioning System). When Pentagon officials first launched the project in 1973, they envisaged huge military advantages by being able to track the location of a submarine or deliver bombs with pinpoint accuracy. Most of us today associate GPS with "turn left in five hundred metres."

When John Daniell invented the 'Daniell Cell' battery in 1836, this first practical battery was used extensively to power the new telegraph networks. I admit there has been slightly more than the blink of an eye but the current range of development of batteries is driven by electric vehicles and mobile phones, neither of which existed in 1836.

Back to my list of brand names. I am writing this as I sit in Prague on a mini-tour of Europe in a way that would not have been possible only a few years ago. When you combine a range of technologies – 4G networks; smartphones; credit cards; GPS and batteries the obvious outcome is…electric scooter hire!

This is a market segment that is currently exploding throughout Europe. The concept seems simple enough – with our modern technology. A company buys a large number of electric scooters and fits them with GPS tracking devices and a QR code. Often without any approvals or permission (legislation is trying to catch up) they then drop them into a city. Paris, for example, has about 15,000 electric scooters. Cologne estimates there will be 40,000 e-scooters by the end of the year.

When anyone wants to use an electric scooter, they download an app, add their credit card details, use location services to locate the nearest available scooter and then scan the QR code with their phone. That scooter is assigned to that person and it is automatically unlocked for use. The initial fee is typically one to two dollars and then a fee in the vicinity of fifteen cents per minute is charged until the ride is finished in the app. As one door closes another opens and there are now people who willingly collect e-scooters at night to charge them at home and deliver them back into the city the next morning for a very modest fee from the e-scooter company.

> During the development of all of the parts of the technology required for this, not one person imagined that their development would lead to this outcome but, once the parts were available, it took some innovative thinking to create an industry.

It hasn't been without controversy though. One pedestrian was recently killed when hit by an electric scooter and several scooter riders have died when tangling with vehicles. Belgium has introduced a speed limit. Sweden has placed a ban of any devices capable of travelling faster than 20kmh. In bicycle friendly Copenhagen, cyclists are now complaining about crowding of their lanes by scooters.

As a solution to urban congestion and climate change, the electric scooter hire model is here to stay.

Tell me which cities in Australia are best placed for an e-scooter invasion at ask@techtalk.digital.

Elementary my dear Google

As I ride my e-scooter through Warsaw and past the bronze statue of Polish astronomer Nicolaus Copernicus, I start to wonder what he would make of one of our modern conveniences. As Alistair Crombie, the Australian science historian, once noted, Copernicus was "the supreme example of a man who revolutionised science by looking at the old facts in a new way."

Copernicus, born 546 years ago, is best remembered for the publication of his book, De revolutionibus orbium coelestium, which proposed that planet Earth did not sit at the centre of the Universe but, in fact, rotated around the Sun.

The Catholic Church banned the book for over two hundred years.

Another famous Polish scientist, Marie Curie, observed that the work of pure science can lead to other benefits. We are seeing that with a humble little tool many of us use every day without a second thought. Google Earth. There is no doubt that the founders of Keyhole Inc., the company eventually purchased by Google to create Google Earth, were visionaries, but in their wildest dreams I don't think they could see that one day their technology would be used to solve cold cases.

What?

The number of mysteries and crimes solved by Google Earth is quite staggering. Remember that Google Earth is nothing like you see in the movies where a user can log onto their PC and, in real time, make a satellite track a car as it speeds across the Nullarbor. Satellite images are usually between one to three years old and Street View updates can be several years out of date. Despite the obvious deficiencies, there are people out there, either bored or with an extreme fascination with Google Earth, that are looking for, well, anything. The most recent case saw the disappearance of William Moldt, 22 years ago, solved by a man browsing Google Earth. He noticed a submerged car beneath the waters in Florida and eventually alerted authorities who discovered William's body.

Back in 2015, a similar situation occurred when David Nile was found in his car in a pond in Michigan. He had disappeared nearly a decade earlier. It is not just disappearing cars that have been spotted though. A criminal drug ring in Switzerland came up with the perfect way to hide their marijuana crop. Surround the drug plantation with a field of corn that grows so high the drugs can't be spotted...except by satellite! A ton of marijuana, literally, was discovered and sixteen people were arrested.

Government authorities are also using the modern tool. In Athens a permit is required to build a swimming pool. The internal records indicated there were only 324 pools throughout the suburbs – which seemed incredibly low. Lacking the manpower to go door to door, the authorities turned to Google Earth – and found more than 16,500 undeclared swimming pools!

The unpaid taxes were not quite enough to reverse the Greek government-debt crisis but they delivered a significant injection to local authorities.

One Italian man reported to authorities that he sold his villa for 280,000 euros and paid the resultant tax. A quick inspection on Google Earth showed that the supposedly small villa was a huge complex in a prime location. Investigators found the selling price was so high that the taxman was owed more than seven million euros!

I love the technology in Google Earth but my use is for typically mundane purposes but if you have the inclination and the time, use the incredible power that is at your fingertips and tell me the most interesting thing you have noticed on Google Earth at ask@techtalk.digital.

The Zombies are coming

Do I need to remove the head? What about sunlight or a cross? No, sorry, that is for the Count. These need a different technique. No stake through the heart but definitely a shotgun aimed at the head. Bludgeoning of the cranium seems like hard work and risky but could also be effective.

What?

Whoops. When I was asked to write about killing zombies, I thought it was a strange topic for a tech article but started my research on Google Scholar (not a lot of peer-reviewed literature I admit).

Now I understand.

Killing zombie apps!

Zombie apps are definitely a problem in the world of smartphones. When you see the claims from a manufacturer of battery life of 25 hours but you struggle to get through a normal day, you often dismiss it as marketing hype (otherwise known as lies). It may not be just over-promising though. You may have some apps sitting in the background that are draining your battery more than expected.

Why are they draining the battery? That brings the next problem with the zombies on your phone. They may be sitting in the background happily communicating with the outside world. Possibly even costing advertisers money for 'displaying' ads on your phone. Do you really care about a large advertiser being charged for ads you never see? Probably not – until you realise it is your data that the app is using to communicate in the background.

Now it is personal.

When your teenage daughter tells you that she definitely could not have used all that data watching *Bachelor* on her mobile, she may be telling the truth (but don't dismiss the *Bachelor* hypothesis completely – we are talking about teenagers here!) While sitting there happily in the background, a zombie app could easily consume tens of gigabytes (GB) in a month.

How is this allowed, you ask? You gave them permission! Remember when you installed that app and it had a bunch of words on screen and you had to accept the conditions to use the app?

What to do?

My first piece of advice is a little boring and will be ignored by most.

Read the conditions on the apps you install. You will be amazed and sometimes amused at what they contain.

Secondly, remove the apps you are no longer using. Make sure you remove them from the phone and your cloud backup as well otherwise they might make a re-appearance. If you haven't used the app for a couple of months, chances are you will not notice its removal. If you really want it back, download it again.

The next step is a little harder. Delete accounts for unused services you signed up for. This has typically been a manual process of visiting the Web site of each of those companies and cancelling your subscription or account. There are too many though! Luckily there is an app for that – or at least Web sites to help.

My last piece of advice is a little scary. On a regular basis, start with a clean install on your phone. Make sure all your important data is backed up and then completely wipe your phone back to factory settings and just setup the accounts and download the apps you know that you need. Not only will it help with the zombies but will probably make your phone run a little faster as well.

On the other hand, now that I have done my research, if you do need some help with the zombie apocalypse, which is surely coming, send an e-mail for some further advice to ask@techtalk.digital.

Riding on the back of sheep

"**W**itchcraft to the ignorant...simple science to the learned." And with that simple statement, science fiction writer, Leigh Brackett, summed up much of the technological environment we see today. The more famous version of that sentiment followed on over thirty years later when, in 1973, Arthur C. Clarke's third law stated that any sufficiently advanced technology is indistinguishable from magic.

Can you imagine taking some of our modern conveniences back only a few decades and showing people of that time. You may very easily have been burned at the stake!

TVs with wireless remote controls only started appearing in the late seventies. Imagine magically making a TV change channel without touching it. Along a similar vein, remote controls for cars didn't start appearing commonly until the nineties.

The first mobile phone call was made in 1973 and commercial TV itself only appeared in this country in 1956. You don't have to go back that far to appear magical by making a glass bulb turn on with the flick of a switch with Tasmania being the first state with a transmitted supply just over one hundred years ago.

There are so many items we take for granted and use every day that are only relatively recent in our history: The Internet; GPS; Streaming, Smartphones...my young teenage daughter rang me last week when I was at 39,000 feet. Knowing that I would incur the wrath of my fellow passengers by taking the call, I assumed it was an emergency as she knew I was between countries. I still think taking a phone call while flying at 0.85 Mach and 12km above the earth is pretty cool.

When I asked about the emergency, there was nothing to report. She was bored and making a call to a plane was no big deal to her!

Without some simple understanding of the technology behind these everyday occurrences, they all appear 'magical'.

It is with that backdrop that I want to look forward. Given the speed of change in the technological world and how silly some technology looks with the advantage of time (I'm looking at *you,* 3D TV) making technology predictions is a dangerous game.

We have only had the Internet in this country for thirty years, but we are on the cusp of having the Internet on, well, everything. With the progression of low power transmissions, advanced batteries and better connectivity, we can connect so many useful items to the Internet. We are already seeing connected cars with the ability to remotely monitor location and speed of a vehicle.

Interesting when you give your teenagers the keys to the family transport but imagine the increased efficiency in an organisation when all fleet vehicles are being monitored in real-time.

Add in a dash of AI and the huge amount of data can be used to track the efficiency of drivers and delivery runs across a company. This is not uncommon already for regional businesses that rely on delivery vehicles to track the exact whereabouts of their entire fleet. Knowledge is power and this delivers a better experience for the clients and is safer for the drivers - this is while we still have drivers but I will leave autonomous cars alone for the moment.

On the other hand, autonomous tractors with some sprinkling of AI are already making a difference to agriculture. Sitting on a tractor for hours on end is boring and falling asleep and running into a fence can be costly. With no traffic to worry about and an open paddock, autonomous tractors are already working the farms in regional areas. The next time you drive on a country road and see a tractor out in the field, have a close look to see if you can spot a driver. When you factor in the ability to monitor the soil and weeds below the tractor and change behaviour accordingly, you start to see some dramatic improvement in efficiency of cropping.

This isn't limited to tractors. Water is a precious resource and fertiliser and chemicals are expensive. Pivot irrigators are now able to install IoT sensors at strategic locations around their irrigated area and the centre pivot used to irrigate an area can then automatically adjust water and additives to specifically what is required at each area. Yield increases of twenty per cent have been achieved with a reduction in costs. One client I know couldn't wait until the mobile reception was good enough at their farm and installed their own farm-wide Wi-Fi system to ensure internal communications were at a good enough level to implement this technology.

This same technology is being developed for broadacre farming although the measuring is more often performed manually due to the larger area being covered. Give it time though.

> As IoT develops and communication links improve, permanently installed IoT devices across the ground will be as common as star pickets across a farm.

And don't think cattle and sheep farmers are being left out of this. As weather patterns change and farmers are trying to extract the most from precious few resources, measuring and monitoring is the new farming mantra. Entire livestock farm management software companies are developing solutions to constant monitoring. Measure what you treasure. Gone are the days of putting your livestock in a large paddock with a dam full of water and coming back when they have fattened up.

Critical measurements are being taken of stock and they are rotated through specific areas of farmland. Water troughs are monitored remotely to ensure a constant supply of water is available. The ultimate aim is to develop the technology far enough to remotely monitor individual animals from a remote location anywhere in the world. In Australia we have land but not a lot of water. If we can be at the forefront of this farming technology revolution, we are well placed to maximise the efficiency of the land despite our lack of water.

It is somewhat obvious but regional communities have dramatically lower population densities compared to metropolitan areas. That is a social strength but it has been a technology weakness. When additional communication links are being built – either wireless or fixed – it is logical that a telco will build infrastructure where population density is greater. Better return on their investment. Logically you would think this would continue with the advent of 5G – but maybe we will be rescued by the sheep!

With 75 million sheep in Australia (and 26 million cattle) if we end up with 5G monitoring devices attached to all of these animals, telecommunication companies may focus their attention on regional areas first!

I may be dreaming but there is logic in the argument.

Much of the world focuses IoT on smart home devices and freight deliveries and bridges and physical items but there is a strong argument that our increased food production should be at the forefront of our technological thinking.

In the same way that the CSIRO has led the world in areas such as Wi-Fi and plastic bank notes, I would be happy with $150 million going to the Australian organisation (rather than NASA) to develop world-leading technologies in IoT for the improvement of product yields.

For the supporters of the CSIRO, that decision would be indistinguishable from magic!

Tell me where I have missed the mark with regional IoT at ask@techtalk.digital.

SPOOF! AND IT'S GONE!

For the sake of the technical discussion, assume for a moment that I particularly dislike the CEO of a fictional company, Blue Widgets Galore (BWG), and also assume I have a very nasty vindictive streak (this is hypothetical remember!).

If I have some time and a little money on my hands, I could quite easily type a letter purporting to be from the CEO. I could create a fake letterhead with all the correct details. On the outside of the envelope, I can print the 'from' address and then send a nasty or damaging letter to anyone I choose. If I wear some gloves when I place the letters in the envelopes and buy my stamps from various locations, I think I could get away with sending this letter to all and sundry and it would appear to come from the CEO. I could tell everyone that the company has gone into liquidation or that I am resigning as CEO or similar malicious messaging.

The point is that we receive a letter in the mail and we trust that it has come from the location stated on the outside of the envelope but there is no actual checking mechanism in place.

The limiting factors are time (to print the letters and put them in envelopes) and money (to pay for the stamps).

Then along came the Internet.

When Internet e-mail was first developed, it was designed for communication amongst a limited number of trusted institutions so security was not high on the priority list. The protocol developed was Simple Mail Transfer Protocol (SMTP) and, as the name suggests, it is relatively simple. Neither the sending or receiving address are checked for authenticity.

Unfortunately, when people with dishonourable motives discovered that they could send a message and appear to be anyone, the previous limitations with the paper version of time and money were suddenly removed. You no longer needed a lot of time or money to pretend to be someone you weren't.

And so e-mail spoofing was born.

> The FBI estimates that e-mail spoofing has amounted to global losses of AU$40 billion over the past three years.

For example, back in 2013, scammers purchased a number of shares in a company and then spoofed the e-mail address of the CEO to send an e-mail to media outlets to inform them of a buyout offer well above the current market price. The offer was reported and shares jumped fifty per cent – allowing the scammers to sell their shares before the scam was realised.

Just last week the CEO of a company in the UK e-mailed his Chief Financial Officer (CFO) to inform him that the £6 million acquisition had been completed and could he please transfer the funds to the attached bank account before the close of day.

Except the CEO didn't send the e-mail. They just waved goodbye to £6 million!

Drop back to a smaller scale. The HR department would not think twice about receiving an e-mail from an employee with a request to change bank account details. The next pay run would proceed as normal – and then the real employee would ask why they weren't paid this week. We have a level of trust with e-mail that would mean we would typically not question the e-mail.

The good news is that if someone spoofs an address and you reply to it, the reply will go back to the correct person. If that person says the original e-mail did not come from them, the alarm bells can start ringing!

Tell me if you have received a spoofed e-mail at ask@techtalk.digital.

WIKIWARS

Since 1768, a small group of people had a very prestigious job. They held a University degree – sometimes plural - and were regarded as the wisest in the land. Parents were filled with pride when they told their friends that one of their children had this particular job.

As with so many careers that change over time, this role in this format has all but vanished, replaced by a technical solution that would have been nigh on impossible to predict in 1768.

The last print edition of the Encyclopædia Britannica was published in 2010 and it spanned 32 volumes and 32,640 pages. 100 editors and 4,000 freelance contributors - all considered experts in their field – were employed to produce that last volume.

Although you can still access Britannica online, ask kids at school today what site they commonly use for school projects, and they will overwhelmingly tell you Wikipedia. For a site to replace such an authoritative tome as Britannica, it must surely be a well-researched and more authoritative publication than the one it replaced with more information and facts and...but no. Forget employing 100 full-time editors who progress through University and apply for a job and collect a wage.

Who edits Wikipedia? Literally anyone can be an editor. And the wage? Zero!

The process is relatively simple. Create a Wikipedia account, familiarise yourself with some policies, click on the Edit button and away you go! That is it! No checking of qualifications. No ensuring you aren't heavily biased in some particular fashion. It is that easy to become a Wikipedian, as editors of Wikipedia are known.

Over 35 million people are registered to edit articles on the English-language version alone with over 300,000 having made more than 10 edits. 150,000 in the last month! It seems like a disaster in the making – but it works. Sure, there are some issues from time to time and they often end up in 'edit wars.' One of the great strengths of Wikipedia is that, unlike the old print versions of Encyclopædia Britannica, when changes occur or mistakes are noted, they can be changed immediately rather than waiting until the next edition is printed – sometimes years away. Its greatest strength is also its greatest weakness – that it can be edited instantly by anyone.

Take, for example, the spelling of the product that soft drink cans are made from. Is it Aluminium or Aluminum? This page has seen 6,000 edits so far. Tropical Storm Zeta formed on 30 December and dissipated on 7 January 2006. Was it the last storm of 2005 or the first for 2006? 3,500 edits have brought no definitive conclusion. Is 'Iron Maiden' better known as a heavy metal band or a German torture device from the Middle Ages? This one is up to 10,000 edits. It is even trickier when it comes to political issues.

In Hong Kong do we have protesters or rioters? We should check that on Wikipedia.

Well, it depends exactly when you check it. With 65 changes of the terminology in a single day, it was hard for the Wikipedians to agree.

As much as some may disregard the quality of Wikipedia and not agree that the masses will ultimately deliver well-researched material, consider an entry in the first edition of Encyclopædia Britannica. The entry for California stated that it "is uncertain whether it be a peninsula or an island." For the researchers in Scotland responsible for the first edition, it was the result of the best research possible at the time.

Tell me what resources you use for research at ask@techtalk.digital.

CEMENTING RENEWABLE ENERGY

Not only does the world we live in change around us but our knowledge of that world also changes. It took until 1543 when Copernicus published his planetary model with the sun at the centre for people to start viewing the planets differently. JJ Thomson proposed a plum pudding model of the atom in 1904. As a physics student, I chuckled at the silly idea as we knew so much better that atoms consisted of protons, neutrons and electrons. A modern physics student would chuckle at that description with quarks and gluons now recognised as a basic part of the atom. The list goes on. We once thought of Pluto as a planet but it turns out it is now only a dwarf planet. We thought our tongue had a taste map with only four tastes (umami is now the fifth recognised taste).

One law of science that hasn't been challenged since it was first proposed in 1850 is the first law of thermodynamics. In short, energy can be transformed from one form to another, but can be neither created nor destroyed.

This principle and concrete blocks are now being used to help solve the problem of storage of power. With solar and wind producing power intermittently, storage of power is going to be more important in our new world free of fossil-fuel generation.

We have seen examples of large-scale battery storage but batteries have two inherent problems. Firstly, they self-discharge at the rate of approximately three per cent per month. Secondly, the materials required for batteries need to be mined and are in increasing demand. Batteries do have the huge advantage that they can be scaled to a variety of sizes and placed where stored power is needed.

Pumped hydro has been a great solution for storage of power in this nation. Over four per cent of our current energy usage firstly goes via pumped hydro. When all of our power was supplied by coal, it was difficult and expensive to turn production up and down in short bursts so excess power was used to pump water up a hill into a dam and when additional power was needed, the water flowed down the hill and turned turbines – with about 80 per cent efficiency.

Snowy 2.0 has a significant capability for more pumped hydro with 350,000MWh of energy storage and 6,100MW of generation capacity. Despite ANU researchers identifying more than 22,000 potential pumped hydro sites across the nation, pumped hydro is expensive to build in areas where the land is flat or there is not enough space to provide a dam and reservoir.

Enter concrete blocks!

A brilliant but incredibly simple idea.

In its prototype form, concrete blocks are lifted by cranes and stacked at height when the grid is supplying excess power, therefore increasing the potential energy of each concrete block. When the grid needs more power – for example at night – the concrete blocks are lowered by the same crane which spins a generator to produce power. The initial towers sit at 120m tall and can store 20-35-80MWh and generate 4-8MW. Significantly smaller than Snowy 2.0 but these towers can be distributed across the network where power is needed and only need a small footprint to construct. Use a denser material than concrete – such as lead or rhodium – and the storage capacity will increase.

Companies will be solving energy problems with innovative ideas such as this while governments of the world are still arguing about energy policies. I can hardly wait to see the next innovation in power storage.

Tell me your brilliant ideas for energy storage at ask@techtalk.digital.

CALLING TIME ON GEORGE

I would argue that I have a desire for extremely accurate data on which to make decisions but some may argue that I have OCD (obsessive-compulsive disorder) tendencies, either way I remember growing up and utilising the services of Australia's version of the 'talking clock'.

Starting in 1953, Australians could ring 1194 (originally B074) and hear the voice of 'George' announce the precise time. The voice was originally provided by Gordon Gow and later by Richard Peach.

I would always like my wristwatch to be accurate to the second so I would ring the service regularly. The cheap digital watches I used growing up were only accurate to within about ten seconds a month so they required regular resetting. I applied the same touch to clocks and later, when installing IT networks, I would set workstation times from the server - but only after I had first set the server time to 1194.

The world moves on so, after sixty-six years of service, the 1194 service has been retired. Before 1953, a call to your local exchange would deliver a live read of the time at the exchange. In 1953, thirty-seven crates arrived from England along with an engineer from the British Post Office to install our first automated speaking clock. Today there are only approximately twenty such services still in operation around the world and despite the fact that one fan in Australia has created a Web site (1194online.com) to keep George announcing the time in Australia, the speaking clock as we know it is gone.

The clock was officially shut down because "it was no longer compatible with new network technologies" but I don't know what that means. I suspect the main reason that it was shut down was that it was a free service that was of little use now that many people use a mobile phone as their accurate timepiece. Our phones, computers and other connected devices can now receive incredibly accurate times automatically – but from where?

Bureau International des Poids et Mesures (BIMP) was setup by the Metre Convention in Paris in 1875 and currently has 102 Member and Associate States and Economies. Coordinated Universal Time (UTC) is set by BIMP based on output from 400 highly precise atomic clocks worldwide.

An atomic clock can keep the time accurate to within one second every 300 million years! It does this by measuring the time it takes a Caesium-133 atom to oscillate exactly 9,192,631,770 times. Seems obvious!

Apart from my OCD, why the need for such accuracy?

Much of the technology around the world would cease to operate without an agreed upon accurate central time. GPS systems rely on triangulation to determine precise positions. The power grid needs to coordinate changes across the network at precise times. The mobile phone network would start to fail if different devices had their time out by just a millionth of a second in a day. Even computers with certain security protocols are requiring more accurate times.

To help our need for accuracy, the Network Time Protocol (NTP) launched in 1985 which allows computers and other devices (for example CCTV equipment) to have accurate times for practical purposes (usually within twenty milliseconds). With the inherent latency of the Internet, it uses a clever synchronisation algorithm to adjust for the round-trip delay. There are pools of servers around the world, available for free, to set your computer to this time.

Maybe it was time for 'George' to be retired, but I will still miss calling 1194.

Tell me if you would still like to call to hear the time at ask@techtalk.digital.

GPS ACCURACY OF TWENTY YEARS

As much as the world of technology has seen some incredible visionaries throughout time, some hindsight short-sightedness demonstrates just how difficult it is to plan for the future. We are all familiar with the Y2k problem, of course, when early programmers deduced that two bytes of data could be saved when writing the date. To any human or computer, it was obvious that if the date man landed on the moon was written as 16071969 or 160769, they were both the same date. At up to US$90 per kilobyte, programmers were encouraged to be more efficient. That was fine until the year 2000 approached and suddenly a date of 020210 could be referring to the year 1910 or 2010!

Despite US$500 billion being spent on solving that issue, we still had some date-related problems after 1 January 2000. Credit card failures; corrupted satellite data; nuclear reactor false alarms; rejection of food shipments; incorrect age-based screening tests for pregnant women and even a video rental late fee of over US$90K!

We don't seem to have too many Y2k problems any more as most systems now use a four-digit date but there are other systems with different date related issues.

GPS springs to mind.

Last week I spoke of the importance of accurate times for different devices across the world. GPS is one of those systems. Einstein's Special Theory of Relativity is used in the calculations for the clock ticks on satellites and reference systems on earth.

The satellites must have an accuracy that is measured in billionths of a second – plus or minus twenty years!

What? Hindsight is a wonderful thing.

GPS week zero started on 6 January 1980. With data at a premium, only ten binary digits were available to count the number of weeks for the date of satellite systems. That meant that 1024 weeks later the system date effectively reset itself to zero. At midnight on 21 August 1999 we had a small taste of what the Y2k problem may bring.

Luckily not a lot of consumer GPS devices existed that had been around for most of those twenty years and systems were generally updated to be aware of this problem. Fast forward another twenty years and on 6 April earlier this year, some airlines had to ground flights while patches were applied. The New York City wireless network crashed and weather balloons and weather buoys were impacted. By this time, the number of consumer GPS devices had increased dramatically – including mobile phones.

Now it might seem as though this problem was all fixed over six months ago but the problem is about to appear again with older iPhones.

Anyone using an iPhone older than an iPhone 6 needs to ensure it has the very latest iOS update. Despite the fact that the GPS rollover date has already occurred, older iPhone models are dancing to a different tune. The same applies for earlier iPad models. There may be other devices as well that have a date offset that may impact the operation.

In short, it is always good advice to keep whatever system you are using patched with the very latest software. Some of my clients over the years have been concerned about applying patches as updates sometimes cause strange behaviour or modify the look and feel of your device but I am firmly in the camp of taking the lower risk of applying the latest patch rather than exposing the possibilities of the problems that are being fixed by the patch.

Tell me if you prefer to patch or cross your fingers and hope at ask@techtalk.digital.

KICK YOURSELF

When it comes to the rich and famous, we are constantly reading about the latest toys and accessories they purchase to live a life that is foreign to the average person. Think of private yachts owned by Steven Spielberg; Paul Allen and Larry Page that feature a full spa, movie theatre, gym and helipad. Think of private jets owned by Oprah; Bill Gates and Tyler Perry that feature a full spa, movie theatre, gym and helipad...OK, maybe not a helipad but you get the idea. The elite in society often show off their wealth with their transport and clothing and handbags - but not their mobile phones.

Apart from some specific security exceptions, in general the phones used by the wider population are the same models used by those we see on the covers of magazines each week. Sure, they may put a diamond-encrusted case on the phone that costs the equivalent of the GDP of a small nation but the phone itself is identical.

And that was never demonstrated more so than last week.

Start with the leader of the free world with a net worth of over US$3 billion. Donald Trump. Now move on to Rudy Giuliani. Giuliani is Trump's personal lawyer and cybersecurity adviser.

One may assume that Giuliani would have access to a secure government issued phone.

Not only does he use a normal, everyday iPhone, but he has trouble remembering his PIN! What does the US President's cybersecurity adviser do when he can't remember his PIN? He keeps trying different PINs until he disables his iPhone.

Next step? Ignore the hundreds of IT staff that work in the Oval Office and visit the local Apple store and ask the teenager without any security clearance for some help.

Just the same as the average iPhone user would do.

In this case a complete reset of the phone was necessary along with the restoration of a backup from iCloud. This is a situation that I am sure Apple staff would see every day – not so much with cybersecurity advisers to the American President but for people walking in off the street.

The lesson from all of this? Be concerned about people in important positions who can't even remember a password? Well…probably…but the real lesson is to ensure that you have a backup of your smartphone.

I have an incredibly complicated practice in relation to backing up your mobile phone. I call it the KYT. Kick Yourself Theory. Imagine one day you are climbing Sydney Harbour Bridge and your phone slips out of your hand into the ocean below.

All data is gone. No contacts, no photos, no text messages. Nothing. How hard do you kick yourself? If the answer is "very" then choose from one of a number of different options to back up your phone on a regular basis. Once upon a time the only method was to plug your phone into a PC and manually backup your device.

Murphy's Law dictated that the day you lost your phone was the day before you were about to do the next backup! Today there are a variety of options to automate backup procedures to store your backup in the cloud on a regular basis.

I normally recommend daily but go back to my KYT. If you lost a week of data is that OK? If so, then backup weekly. Would you kick yourself if you lost data from an hour ago? Then backup hourly. Most importantly, be aware of your backup scenario BEFORE you lose your data.

Tell me how often you perform a phone backup at ask@techtalk.digital.

GOOGLE THAT BRAND NAME

Today I have a big day planned. After a breakfast of yeast extract on toast, I am going to visit a shopping mall and travel down the moving stairway after having purchased my lightweight jacket with both clasp lock zip fasteners and additional hook and loop fasteners. I am going to ride my personal watercraft before throwing a flying disc to my children who will, no doubt, ignore the cold and wear a short-sleeved, collared shirt. I will need to pack some facial tissues and adhesive bandages for the outing plus bring a vacuum flask for morning tea. I wonder if I should just cover my children in petroleum jelly and wrap them in air bubble packaging?

This statement has been uttered...never...because certain brand names are now synonymous with the product. If you replaced descriptions above with Vegemite; Escalator; Windbreaker; Zipper; Velcro; Jet Ski; Frisbee; Polo; Kleenex; Band-Aid; Thermos; Vaseline and Bubble Wrap the statement makes more sense.

Google was genericised in an incredibly short timeframe.

After being founded in 1998, it was only 2002 when the American Dialect Society named google, the verb, the most useful word of 2002.

No longer did we search for something on the Internet but we had started 'googling' search terms. The first usage in American television was also in 2002 when Willow asked Buffy (the Vampire Slayer) if she had "googled her yet." The Oxford English Dictionary added the term in 2006. Little did it matter that there were dozens of search engines used by the public. Google had attracted enough market share that people started to google everything. Large organisations stopped putting their full Web address on advertisements and instead started to tag ads with "just google Product X" as it was easier to remember than a Fully Qualified Domain Name (FQDN).

Suddenly scammers smelt an opportunity.

If a scammer setup a Web site to look identical to, say, a banking site then the scammer could rely on people googling the banking site to direct some traffic to their site. Throw in a few Google ads and suddenly the fake banking site appears at the top of the results. Once you type in your details to the fake site the scammer receives your details and bada bing bada boom they are draining your bank account.

The opportunities were not just in relation to bank sites. Government Web sites – in particular the tax office – are common targets of fake Web sites. Download software to your computer that stops viruses or performs a useful function and, better still, is free, sounds fantastic...until you realise that you have just downloaded a virus to perform malicious activities on your PC.

Perhaps the scariest googling of all occurs with what is often called Dr Google. 80 per cent of Internet users said they searched for a health-related topic online. Ignoring the qualifications issue and your specific circumstances – both red flags – it seems uncanny that many of these medical advice sites seem to ultimately direct you to a 'miracle cure' for only one small fee. When the cure is the same despite the fact that you may have a skin condition or high blood pressure or headaches, you should be very sceptical.

It may sound like contrary advice for a technology column, but as much as googling is a term that we use so often in our daily lives, my advice is not to google everything. There are still some instances when typing in a FQDN, albeit tedious, is a much safer solution.

Tell me if you have ever been directed to a fake Web site at ask@techtalk.digital.

POWERFUL IMAGERY

I **had a trip down to Bega and Eden on the picturesque Sapphire Coast this week – and apart from a variety of beautiful views I came across an incredibly unusual sight.**

I saw a person taking photos...with a camera!

Sure - people still do use cameras to take photos but I admit to being a little surprised when I see something other than a smartphone used for photography.

And with good reason.

Of the estimated one trillion photos taken across the world last year, eighty-nine per cent were taken on a mobile phone.

Many people now cite the camera specifications as the main reason for their specific choice of mobile phone!

A camera on a device that is conveniently sized and is always with you sounds great but how do the pictures compare to a 'real' camera?

With the increasing power of a modern smartphone, they are actually pretty good. That last sentence doesn't make a lot of sense. Surely it is all about the lens and the sensor – not the processing power of the phone? Well...yes and no.

Firstly, we should look at the power of a modern smartphone. I am going to oversimplify a comparison here but the latest smartphones have processors that are capable of approximately one trillion operations per second. In 1969 NASA used five IBM System/360 Model 75 mainframe computers to put man on the moon. A Model 75 would set you back a cool US$28 million in current terms and was the size of a small car. It could perform one million operations per second – so a modern smartphone is one million times more powerful. Even IBM's famous Deep Blue, the first supercomputer to defeat a reigning world champion in a chess match, could only perform one million operations per second. So a modern smartphone has some power.

The sensor where the image is captured is surely incredibly important? A modern high-end DSLR camera would typically have a full-frame image sensor that measures 36mm by 24mm. A high-end smartphone camera may have a 1/1.7" sensor that measures 7.6mm by 5.7mm or twenty times smaller than a DSLR. The actual lens is sized accordingly. Logically the increased sensor size gives you a greater surface area and therefore a better photo.

That is largely correct but the incredible power of a smartphone is delivering surprisingly good results with small sensors. Algorithms using that power are further enhancing those images such that a simple smartphone camera is delivering pictures that rival professional cameras. An inherent weakness of a small sensor is low-light photography but, as one example, computational photography can create a merged image of multiple photos of different brightness to create an enhanced image. Similarly with a panorama photograph where an unsteady handheld phone can be panned across a scene while the phone takes multiple images and stitches them together. New algorithms being developed will remove water from underwater images so, rather than a dull blue tint that is often associated with underwater photographs, the objects will appear as if they were sitting on land.

I haven't even touched on the size of the image with most phones delivering image sizes of twelve to twenty megapixels but some as high as forty-eight – but don't make a rookie mistake and use megapixels as your only rating of a camera.

With advanced bokeh effect and different cameras for different zoom levels and filters and effects and so much more, the hardest aspect of modern smartphone photography is deciding which features to use!

Tell me if you now take your photos on a smartphone at ask@techtalk.digital.

Slippery poo

Most people see the name of my column, *'Tech Talk'*, and assume I talk about technology. Correct assumption. Where it starts to become more complicated is in the definition of technology. For most people the modern assumption of technology – in some part confirmed by the topics I discuss – would be that technology needs to have some component of electronics associated with it. Computers; mobile phones; electric cars; renewable power; the Internet; etc.

These are all topics I have written about in recent months and all topics that involve the flow of an electric charge. With this in mind, you may well say that this column should more accurately be called 'Electric Talk'.

Apart from the fact that it doesn't involve alliteration, I would prefer to go back to the definition of technology. It comes from two Greek words, tekhnê meaning art, skill or craft and logia meaning words. So technology can correctly be used as a term to encompass any tools or devices or even techniques that are used to help us.

By that definition, stone tools could easily be classified as an early example of 'modern technology'.

There is a reason for delving into the history of the word. My topic today does not involve the movement of electric charge but instead the movement of something much easier to see.

Poo.

With many parts of Australia currently in the grip of possibly the worst drought on record, a variety of techniques are being used to reduce water usage. The toilet once was a source of a huge amount of water usage with early toilets using over twenty litres per flush. In 1980 an Australian company designed a dual flush toilet which reduced the volume to eleven and six litres per flush and a re-design in 1994 cut that back to six and three. With all those advances in toilet technology, the average household still uses up to twelve per cent of its water on toilet flushes.

But that may soon change.

Scientists at Penn State University have created an ultra-slippery toilet coating that is claimed to reduce the amount of water required to flush solids by ninety per cent. And here I was thinking that porcelain was already pretty smooth!

The two-step polymer spray takes approximately five minutes to cure and the early prototype is useful for up to five hundred flushes. The surface ends up being more slippery than the well-known non-stick substance, Teflon.

As with so many advances in technology, researchers looked to nature for inspiration. The carnivorous Pitcher plant has a rough surface that becomes lubricated when it rains so that insects slide inside to be digested. Just like the plant, the design uses two separate coatings which creates a combination of roughness and lubrication. When the first coating is sprayed on the ceramic surface it covers the surface in tiny polymer 'hairs' that attach to the surface. The second spray coats the microscopic hairs with lubrication.

This spray has come about as part of the 'Reinvent the Toilet Challenge', a concept to bring sustainable sanitation solutions to the almost three billion people worldwide who don't have access to safe, affordable sanitation. This challenge is backed by the Bill and Melinda Gates Foundation – with Bill having made his money, of course, in a more traditional form of modern technology.

As much as I am impressed by the reduction in water usage there is an additional side-benefit. The super-smooth surface also repels bacteria which reduces odours! For some this may be reason enough.

Tell me if you think this article is pure excrement (pun intended) at ask@techtalk.digital.

TYSM 4 READING BRB NEXT WEEK

What would your answer be if I asked you for the most commonly used data service in the world? Just a subtle hint - I have used exactly 160 characters thus far.

It was twenty-seven years ago that the first text message was sent.

Now text messaging – or more accurately Short Message Service messaging – is the most popular data service in the world. With good reason. Text messages have an open rate of 98 per cent and a response rate of 45 per cent. By way of comparison, e-mail has only a 20 per cent open rate and 6 per cent response rate.

It is a large number – but probably not that surprising then – that there are currently 23 billion text messages sent on a daily basis across the world.

It is fair to say that on 3 December 1992 when Neil Papworth sent that first text message, "Merry Christmas", to Richard Jarvis, a director of Vodafone UK, that neither Neil or Richard would have possibly predicted how invasive text messages would become in our daily lives.

In the initial design of mobile phone networks background data is exchanged between a phone and a mobile phone tower using the control channel. It was in later development that the idea of using this control channel to exchange a text message was investigated – hence the limitation of 160 characters.

One of the huge advantages of text messages is that you can send and receive a text message to any other person regardless of their handset or carrier or location.

Other message services allow more content and may use different tools but you are relying on the receiver to have the same message service. The latest smartphone that has every feature imaginable can still send a text message to your luddite friend that is using a mobile from the last decade.

That is one of the reasons that so much development has occurred around text messages. Organisations from medical professionals to hairdressers and everywhere in between now send text messages as reminders to clients for appointments – with the ability to confirm the appointment in reply. Online contests are also a popular usage of text messages – but just be wary of the potentially hidden costs. Many shows have been created around the concept of the public voting for their favoured contestant using real-time text messaging. The TV executives love it as it allows them to gain instant feedback on viewer numbers and engagement and the viewers feel involved in the contest.

As we move into a world of smart home automation, text messages are being used to interact with various devices around the home. Many devices use SMS for notification of events (front door was opened) or allow you to trigger events (open the front gate). Smart cars can also be used in a similar way with some cars able to text the owner if a car goes over a certain speed or outside a pre-set range. Fantastic when you give your car to your teenage child (not speaking from experience here). These text messages are given the term 'tattle text' for obvious reasons.

If you fancy yourself as a fast texter, type the following phrase on a touch-screen phone with no predictive features turned on: "The razor-toothed piranhas of the genera Serrasalmus and Pygocentrus are the most ferocious freshwater fish in the world. In reality they seldom attack a human."

If you can do it in under 17 seconds, there is place waiting in the Guinness World Records for you.

TYSM for reading BRB next week C U L8R MSG me @ ask@techtalk.digital.

POUCHING YOUR PHONE

A friend of mine once told me that a phone call is a demand for a meeting without an appointment. Whilst I don't necessarily agree with the attitude from a customer service perspective, I do understand the sentiment. Until 1876, our society existed on the basis of physical interactions or the laborious method of communication that seems non-existent today – letters. Once the telephone became entrenched in our society, communication became so much quicker.

When we think of our society today, we have gone a step forward – or some may say backward – with the ability to access almost limitless information in the palm of our hands. The ongoing debate is whether the incessant use of a smartphone is beneficial to individuals and society.

I am constantly intrigued by the number of new market segments that develop out of our constantly changing world and over-use of smartphones has spawned another business segment.

Enter phone pouches.

Several years ago, Graham Dugoni felt frustrated that people at concerts he was attending seemed more absorbed with their phones than the concert. Yondr was subsequently founded and created an entirely new market segment.

At first the concept seems unnecessary – until you realise that mobile phone use is an addiction to some people. As you enter a concert supported by Yondr, you are handed a pouch and you place your phone inside. The case is electronically latched and cannot be unlatched until you leave the concert and open the pouch at an unlocking station in the foyer. You maintain possession of your phone at all times therefore removing the fear of theft - but you simply can't use it. As much as it would seem simpler to ask everyone to turn their phones off for the concert, we have all been at events when this request has been made and not a single member of the audience reaches for their phone to turn it off!

It would seem the only way to have a phone free concert is to enforce it - although there is some irony in the fact that a technology solution is being used to address the reliance on technology!

Although the concept started with concerts it has now been extended to a number of other areas. I remember my first experience of entering the public gallery of our Federal Parliament and being asked to hand my phone to the door attendant. It made me feel very uncomfortable handing my phone to a stranger. The Yondr concept is being applied in many situations such as this – parliaments, court rooms, weddings, events and even schools are now using this same concept.

The idea of a phone pouch at school is an interesting one. If the number of schools in the nation is n, the number of mobile phone policies across the nation seems to be approximately n + 1. Psychologists, educators, parents, kids, my friend who doesn't like phone calls...they all have a strong and defensible opinion on mobile phones in schools. One school in Australia has just started with a test program of using pouches and already they have noticed more activity and playground noise during recess and lunch. This is only anecdotal at this stage and I am quite certain that a significant amount of research will continue to be conducted.

What I need next is a home version of the pouch so I can sit down and talk to my children at the dinner table rather than phones being used during dinner.

Unfortunately, the worst offender is me!

Let me know what you think of the idea of phone pouches at ask@techtalk.digital.

WHO IS WATCHING YOU?

A s our family gathers together during the holidays, it is inevitable there will be a point in time when we start reminiscing about our lives together. We pull out funny stories from our past with maybe a little embellishment on each retelling.

One chestnut that always features in our ruminating is the time when one of our teenage children snuck out to a party late at night. When my wife and I were woken by our dog, a quick inspection of the house found no intruder, but a decision to check our CCTV footage delivered immediate results.

Said teenager jumped out of bed – fully clothed – and burst into tears with a full admission about the nocturnal activities!

It wasn't that long ago that the thought of CCTV in your home was unheard of – but the development of CCTV now sees (only one pun this week, I promise) over 400 million cameras installed across the world.

CCTV has a long history. In 1927, the Moscow Kremlin installed a system to allow Stalin to monitor approaching visitors. In 1942 in Peenemünde, Germany, another early system was installed to allow Hitler to observe the launch of V-2 rockets.

When CCTV moved from monitoring to recording and monitoring, the value of CCTV increased dramatically. The first systems involved manual reel-to-reel media but with the advent of VCR during the seventies, tapes could now easily be inserted and swapped or, better still, rewound and recorded over again. Outside the banks and casinos and streets of New York, it took the development of digital multiplexing during the nineties for CCTV to really become mainstream. Digital Video Recorders (DVR) were introduced at the beginning of this century which allowed recording of footage on hard drives for even greater flexibility and faster retrieval.

The current crop of cameras are high-quality digital IP cameras that process the signal at the camera itself and then send the image for viewing and recording via ethernet cables. The data can be stored and viewed via a Network Video Recorder (NVR) or...the cloud.

The last component is the area with the greatest excitement...and concern. An NVR uses hard drives to store data in the same way as a DVR but with the data being in a digital format, it makes it very easy to view the data via the Internet. Some cameras even bypass a local NVR altogether and record the data directly to the cloud. That sounds fantastic – a thief wanting to destroy evidence at a crime scene can potentially find a local storage device but it is harder when the data is in the cloud. Or maybe not...

A family in the US recently installed an IP camera to provide additional safety for their 8-year-old daughter. Whilst innocently playing in the same room as the camera a few days after purchase, a stranger struck up a conversation with the girl – via the camera. Not only could they see everything the camera could see, but they could speak with the girl. In this case the girl was aware enough to report the incident to her parents but...it could have been much worse. The provider of the camera emphasised the importance of secure passwords and two-factor authentication and I concur with those statements. The parents admitted they had a very lax attitude to security – which changed very quickly after this incident.

Whether you like the idea of CCTV or not, the reality is that you are probably being recorded by multiple cameras every single day as you go about your daily life.

Tell me your thoughts on CCTV in our society at ask@techtalk.digital.

LOOKING BACKWARDS

This Christmas, I promised myself, I definitely wouldn't overindulge and then collapse in front of the TV and start pumping out zeds. But with a beautiful meal prepared and family gathered, the inevitable happened and before I knew it, I looked like Homer Simpson on the couch complete with dribble running from the corner of my mouth.

In much the same way I promised myself this year that I wouldn't do a 'Year in Review' column but the inexorable draw of the end of year retrospective column has consumed me and, try as I may, I can't help but look back over some of the technology from this year.

For the record, this is NOT an end of decade review. In the same way that all those parties on 31 December 1999 to celebrate the end of the century got it wrong, the end of this decade doesn't occur until the end of 2020. Phew! I got that off my chest!

Despite the unfounded health concerns, the 5G network is one of my highlights. Although technically Telstra switched on 5G in ten cities last year, this year has seen an increased adoption of 5G with over twenty-five cities across the nation having access to 5G and, just as importantly, several devices available that support 5G, including a mobile broadband Wi-Fi device. Australia is one of only ten countries with 5G coverage of any note.

With promises of greater speed, more capacity and lower latency, 5G will enable the unlocking of a huge number of other technology advances including the world of IoT (Internet of Things).

The wearables market is the next area that saw some giant leaps in 2019. Who needs a doctor when you have a watch to assess and monitor your personal health? Watches now allow continuous heart rate monitoring; ECG; sleep tracking; cardio fitness measuring; menstrual cycle tracking and a great feature that I cannot convince my children to use. Noise monitoring to help the wearer of a watch determine when a concert may be damaging your hearing (you would think the ringing sound for two days following a concert would be a subtle hint!) With international emergency calling with fall detection, wearers really feel safer and healthier with their latest wearable.

With the health of the planet in mind, electric cars have been around for some time but this year added some interesting twists. Lotus introduced a fully electric supercar; Lamborghini produced a hybrid production car and the bastion of petrol-heads, Harley Davidson, added electric motors to their mix. The 2020 Olympics will see autonomous electric vehicles being used to transport athletes. And the darling of the EV industry, Tesla, introduced their Cybertruck - with mixed reviews.

There were some negatives as well. Though technically the Cambridge Analytica scandal was a 2018 news story, the fallout continued in 2019 with the Federal Trade Commission issuing a US$5 billion fine and privacy concerns given as the main reason for the loss of 15 million Facebook users.

Unfortunately, I can't have a year in review without mention of the Samsung Galaxy Fold, the revolutionary folding mobile phone that threw in other features for free such as delamination and fragility. Foldable phones will be in our future, just not part of 2019.

The Boeing 737 MAX also shows how we can sometimes rely on technology too much. Complete trust in an overaggressive automation system was given as a direct contributor to two crashes with significant loss of life.

While I prepare for some snoozing during the Boxing Day Test, tell me your biggest technology moments of 2019 at ask@techtalk.digital.

Mathew is rare in the technology journalist space. He doesn't just talk the talk – he walks the walk. His IT career started as a 13-year-old when his school made a progressive decision at the time and bought two devices that teachers had been talking about at conferences. Computers. Apple computers to be exact.

To say Mathew took to them like a duck to water is an understatement. Within days, Mathew was writing programs for them and quickly secured the job of entering the school student body into a database (possibly because most of the teachers didn't know how to turn on a computer). From this time on, Mathew realised that he had found his passion – technology.

Mathew started his first micro-business at the age of 12 and started his first IT company on 4 December 1989 as a brash and naïve 21-year-old. After the success of this business, Mathew went on to spin-off more technology start-ups. ComStall – a communication installation and cabling business. AXXIS ComWorld – a mobile phone retail business. 3ti - the technology training institute. Small Business Ru!es – an IT and business consultancy service. He also developed the SLAM DVD which helped IT businesses in 17 countries transform from a break/fix model to a Managed Services model. This was many years before the MSP concept became the only way to run an IT business.

These businesses collected a plethora of awards while under Mathew's guidance. The Microsoft Worldwide Partner of the Year award; the Microsoft Asia Pacific Certified Partner award; the Australian Business Council award for Innovation; the Australian Small Business Champions award for the best IT business in Australia; the Microsoft Australia/New Zealand Platform Partner of the Year; The Australian Small Business Champions award for the best regional business in NSW; the Telstra Partner award for Most Improved Independent Store in Australia; the MSPMentor USA award as a Top 100 Worldwide IT businesses and a Gold Rhino award as the best overall business in the region as well as multiple Silver Rhino awards for service; employment practices; presentation and marketing. Mathew's businesses have collected almost thirty major awards over the last two decades.

Mathew has also collected many personal accolades. He was showcased as the only Australian in the US published book *'Top 15 Successful SMB Consultants Worldwide'* and featured in the life advice book *'What I wish I knew at eighteen'*. He was named by MSPmentor as one of the Top 250 MSP individuals worldwide. He has won the Consensus IT Writers Award as the Most Entertaining IT Writer in Australia and a previous book of his won a bronze medal in the Axiom Worldwide book awards.

Mathew's business success gives him a 'hands on' perspective to take his knowledge to an international audience – which is exactly what he has been doing since 2004. That was the time when he first started speaking at conferences across the world for organisations such as Microsoft, Intel, Telecom, Kaseya and Autotask and has performed session and keynote presentations here in Australia as well as through locations such as Europe, New Zealand, South Africa, Sri Lanka, the United Arab Emirates, and the United States.

Not only does Mathew perform presentations but writes and talks technology. He has a current radio segment called *'Tech Talk'* where he updates audiences on a weekly basis for three different radio stations. He writes a weekly column, again called *'Tech Talk'*, which appears in 140 mastheads across Australia and has written for a variety of publications such as Computer Reseller News with *'From the Coal Face'* and APC Magazine and TechGenix in the USA. Mathew is a highly respected IT entrepreneur and commentator in the Australian IT and SMB space.

When he does find some spare time, he races Mountain Bikes with his local club; writes and recites poetry and spends time with his wife of twenty-five years and four children. He is also a dedicated community member having recently spent five years as Mayor of his city and now regularly organises fundraisers for important charities and facilities including securing several Guinness World Records as part of those events. He is an active member of his local Rotary Club and enjoys his Mensa membership.

'Tech Talk 2020' is his fourth published work. You can contact Mathew at md@mathewdickerson.com.

DICKERSON

TECHNOLOGIST · FUTURIST
SPEAKER · AUTHOR · POET

www.ingramcontent.com/pod-product-compliance
Lightning Source LLC
Chambersburg PA
CBHW051413200326
41520CB00023B/7211